Praise for *Atmosphere of Hope*

"Best-selling Australian author Tim Flannery counsels cautious optimism by showing how the millions of small actions taken by individuals are driving down oil consumption and points out how new 'Third Way' carbon-capture technologies promise to reduce emissions and create massive economic opportunities."

—*National Geographic*

"Certain people rise above the crowd. They choose to use their extraordinary talent and intelligence to make a difference. Scientist, scholar and activist Tim Flannery is one of those rare people . . . *Atmosphere of Hope* is a brilliant examination of where we are with climate change and where we might be able to go."

—*National Observer* (Vancouver)

"The book does a remarkably good job of arguing that there is still hope for averting catastrophic climate change . . . [Flannery] fully acknowledges the steep challenges and serious obstacles we face. So when he affirms that a path to averting catastrophic climate change remains in place, we know the conclusion is not reached capriciously."

—*Los Angeles Review of Books*

Praise for Tim Flannery's *The Weather Makers*

"An authoritative, scientifically accurate book on global warming that sparkles with life, clarity and intelligence."

—*Washington Post*

"At last, here is a clear and readable account of one of the most important but controversial issues facing everyone in the world today. If you are not already addicted to Tim Flannery's writing, discover him now."

—Jared Diamond

ATMOSPHERE OF HOPE

Other books by Tim Flannery

The Weather Makers

Mammals of New Guinea

Tree Kangaroos
with R. Martin, P. Schouten, and A. Szalay

Possums of the World
with P. Schouten

Mammals of the South West Pacific and Moluccan Islands
Watkin Tench, 1788 (ed.)

The Life and Adventures of John Nicol, Mariner (ed.)

Throwim Way Leg

The Birth of Sydney (ed.)

Terra Australis (ed.)

The Eternal Frontier

The Explorers

A Gap in Nature with P. Schouten

Astonishing Animals with P. Schouten

Chasing Kangaroos

The Future Eaters

Now or Never

Here on Earth

Among the Islands

An Explorer's Notebook

ATMOSPHERE OF HOPE

SEARCHING FOR SOLUTIONS TO THE CLIMATE CRISIS

TIM FLANNERY

Grove Press
New York

Versions of the Afterword originally appeared in the *Sydney Morning Herald* ("The Great Barrier Reef is losing its adjective and it's our fault") and in the *New Daily* ("Tim Flannery: the biggest surprise from the Paris Climate Meeting")

First published in Australia in 2015 by The Text Publishing Company

First published by Grove Atlantic, October 2015

FIRST PAPERBACK EDITION, October 2016

Printed in the United States of America

ISBN 978-0-8021-2565-1
eISBN 978-0-8021-9092-5

Grove Press
an imprint of Grove Atlantic
154 West 14th Street
New York, NY 10011

Distributed by Publishers Group West

groveatlantic.com

16 17 18 19 10 9 8 7 6 5 4 3 2 1

Contents

FOR ROB PURVES,
WHOSE LIFELONG DEDICATION TO THE
ENVIRONMENT HAS CHANGED
THE WORLD

Introduction

> What is the use of having developed a science well enough to make predictions if, in the end, all we're willing to do is stand around and wait for them to come true?
>
> F. SHERWOOD ROWLAND, NOBEL LAUREATE FOR HIS WORK ON THE OZONE HOLE

READERS of this book will discover that we are already living in the climate future. We are already confronting dire scenarios—the melting of the polar ice, the profound degradation of the Great Barrier Reef, the displacement of people in coastal cities due to extreme weather.

And yet, even though I have tried to explain the terrible outcomes that await us if we do nothing about carbon pollution, I have called this book *Atmosphere of Hope*. That may seem to be a strange starting position from which to argue for renewed optimism. But if we are to have real hope, we must first accept reality. We must cut through the dense and complex debates about climate that leave many feeling lost and paralysed.

This book describes in plain terms our climate predicament, but it also brings news of exciting tools in the making that could help us avoid a climatic disaster.

In December 2015 the nations of the world will gather in Paris to try to forge a climate treaty designed to give us a fighting chance to limit global warming to 2°C, widely regarded as the safe upper limit if we are to avoid catastrophe. If we succeed at Paris in forging a new era of international political cooperation in the fight against a warming planet, it is possible that the next decade will astonish us with the solutions that we discover to safeguard our planet for our grandchildren and their grandchildren. We will quite literally create an atmosphere of hope.

Let's try to understand the dimensions of the problem. Projections indicate that in 2014 we humans released a record 40 gigatonnes of CO_2 into the atmosphere,[1] of which 32.2 gigatonnes came from the burning of fossil fuels for energy (mostly electricity and transport).[2] CO_2 is only one, albeit an important one, of over 30 known greenhouse gases. If we add together all human-related greenhouse gas emissions and express them in terms of CO_2's warming potential, the figure at the end of 2009 was 49.5 gigatonnes of 'CO_2 equivalent'.[3] In other words, the warming potential of nearly 50 gigatonnes of CO_2.[4]

A gigatonne is a billion tonnes—a number with nine zeros after it. Even measured against the world as a whole, a gigatonne of CO_2 is large. How large becomes clear if we consider what would be required to take four gigatonnes of CO_2 out of the air. All of the world's agriculture and forestry waste, and the biomass from 100,000 square kilometres of sugarcane, would need to be turned into biochar to do that. And four gigatonnes of CO_2 is only a tenth of our annual CO_2 pollution stream. Alternatively, if we planted forests over an area the size of Australia, or the contiguous 48 states of the USA—assuming that we could plant an area about the size of New York State each year—we'd only reduce our pollution stream annually by a similar amount, averaged over a 50-year period.

No matter how you measure it, our climate problem is now gargantuan, and it has grown at a far faster rate than almost anyone imagined it would just a decade ago. We should be focusing on reducing emissions by the gigatonne. Frustratingly, the objective of the political negotiations is expressed in degrees Celsius rather than gigatonnes of carbon. It is widely anticipated, however, that an agreement at Paris will put the world on a carbon budget that will give us a 50–50 chance of keeping average global temperatures no more than 2°C warmer than they were prior to the Industrial Revolution. This would be a huge improvement over our current worst-case scenario path. And it's possible that, if we can reach agreement, the deal might be subsequently improved, perhaps by providing a five-year review of the target. It is important to understand that such an agreement on a warming limit of 2°C would constitute a massive international breakthrough, but it nonetheless does stake our future on the toss of a coin. We ought not contemplate a future in which we accept that there is still a significant chance that the Great Barrier Reef will die, that sea levels will rise rapidly, and that great losses in biodiversity will be inflicted—all of which are likely to happen in a world 2°C warmer than the pre-industrial average. Global agreement in Paris is necessary but is in itself not enough. Instead we need a more ambitious work program, with more tools to employ than we now have. Let's try to put the Paris meeting in context.

The Paris agreement won't require action until 2020. Perhaps the slow start is inevitable. When Al Gore released *An Inconvenient Truth* and I published *The Weather Makers* a decade ago, climate change was thought of as a hypothetical issue by most people. You had to understand complex graphs and computer models to grasp it. As of 2014 most people have experienced enough extreme and record-breaking weather to know that climate change is not only real but also threatening to their health, livelihoods and security. Across the US, for example, average temperatures have risen by up to 1°C, and Americans know from bitter experience that with the rising average come savage extremes. In the next few decades temperatures across the US would rise by half a degree Celsius even if all global emissions of greenhouse gases had stopped instantly in late 2014. We are late—very late indeed—in acting to secure our future.

We've not lacked warnings about the seriousness of the situation. Ten years ago I published *The Weather Makers*, a bestselling book about climate change. It was just one voice among many, the first being Bill McKibben's *The End of Nature* in 1989. Ever since, the warnings have come thick and fast. Among the most important have been those from scientists. They have framed a global carbon budget that shows that action deferred until after 2030 may be ineffective in helping us avoid severe, ongoing impacts from out-of-control climate change. The work of the International Energy Agency (IEA), a Paris-based intergovernmental organisation that provides statistics on energy-related matters, has been unequivocal. In 2012 the IEA announced that by 2017 humanity would be committed to warming 2°C above the pre-industrial average if investment plans in energy did not alter.[5]

I know from direct experience that the warnings have had

some effect. *The Weather Makers* helped alert Sir Richard Branson to the dangers of climate change, and he recommended the book to Arnold Schwarzenegger, then governor of California, who helped lay the groundwork for the state's clean-energy boom and carbon-trading scheme. Branson also established the Virgin Earth Challenge—a search for solutions with the potential to draw CO_2 out of the atmosphere on a large scale—as well as the Carbon War Room to encourage market-driven gigatonne-scale climate mitigation. The Carbon War Room is now presided over by former Costa Rican president José Maria Figueres, and it recently joined forces with the Rocky Mountain Institute. Gordon Campbell, then premier of British Columbia, told me that he introduced the province's carbon tax after reading *The Weather Makers*; and Professor Zhou Ji, chairman of the Chinese Academy of Engineering, said that the book opened his eyes to the extent of the climate problem. *The Weather Makers* was translated into 23 languages and read by millions, many of whom took their own, individual actions to help reduce carbon emissions.

Active leadership in business and government has also driven markets: over the last few years the growth of solar and wind energy has been spectacular, and electric vehicles are now looking like the obvious future of road transport. These are promising beginnings, but no more than that, for our emissions continue to grow. As a result, the atmospheric concentration of CO_2 reached 400 parts per million (ppm) in mid-2013, the first time it had done so for millions of years. Because of seasonal variations it soon dropped back. But in a few years CO_2 concentrations will be permanently above the 400-ppm mark.

In all the history of our planet, geologists can find no time when atmospheric CO_2 concentrations have risen so quickly. And there is no doubt that the rise has been caused by us, for we can measure how much fossil fuel we burn, and we know precisely how much CO_2 is thereby created. We know also that there can be only one successful outcome. Before 2020 we must achieve an absolute decrease in annual global carbon emissions—and that means reducing the amount of fossil fuel burned. Astonishingly, we now have the first intimation that this might be possible. On 13 March 2015 the IEA announced that the growth in CO_2 emissions from fossil-fuel energy sources had 'stalled' at 32.3 gigatonnes—the same figure as reported for the previous year. This was the first time, the IEA said, that CO_2 emissions had not increased at a time of economic growth.[6]

Even if 2014 marks the peak year for global CO_2 emissions from the energy sector (and it is far too early to tell whether emissions will fall or rise in future), the battle for a stable climate will be far from over. It's clear that we will need to achieve massive cuts in emissions between 2020 and 2030, and to eliminate greenhouse gas emissions from the burning of fossil fuels by 2050. Addressing climate change will thus define the lives of generations.

One thing that gives me hope for success is the increasing empowerment of the individual. When I wrote *The Weather Makers* the best I could suggest for those hoping to take individual action was changing light bulbs and other such efficiency measures, or becoming involved in our encumbered political systems. But digital connectedness has brought new opportunities: for divestment, effective dissent, encouraging uptake of new technologies, and for legal action.

As I researched this book I came across the most astonishing solutions to our climate problem. None is a silver bullet, but many have the potential to make big contributions. Their diversity, in nature and in the regions they are taking place in, is heartening. Even in poor nations people are working on effective and innovative solutions. It does seem now that much of the world is finally beginning to act.

Our inconsistent political approach has also yielded some surprising results. From Germany to the US and Japan, the developed countries that acted under the Kyoto Protocol, or in other ways, have created strong economic growth while sharply reducing emissions. Breaking the link between prosperity and pollution and creating new links between wealth and clean power are the best signs yet that we might keep ourselves safe. Just how and why this is happening, and where hope springs in the rest of the world, is a big part of this book.

The Copenhagen meeting in 2009 was derided by the climate-change sceptics and polluters as an utter failure—indeed they mocked it as the final failure of efforts to create an organised global response to planetary warming. Yet it was at Copenhagen that President Barack Obama, along with the leaders of China, India, Brazil and South Africa, brokered a deal in which countries set their own targets for emission reduction over the critical decade. The Copenhagen Accord, as the one-page agreement became known, set countries free to tackle the problem on their own terms and in ways that suited their individual economies. And because countries set their own targets, failure to achieve them would be a sign of incompetence and cause loss of confidence in governments.

More and more countries signed up in the months following the Copenhagen meeting, and today the Copenhagen Accord is the basis of global climate action.

The trouble is that the cumulative pledges to cut carbon pollution made under the Copenhagen Accord would, if all fulfilled, deliver only half the reductions required to avoid more than 2°C of warming. Despite this, the Copenhagen Accord remains our best proof that we can act globally to meet the climate crisis. Yet the question looms: can we improve our performance? Will we, indeed, perform the perfect three-point turn that's required now to avoid disaster and steer humanity to a safe climatic future? I do not believe that we are, for all of our ingenuity, capable of perfection. We therefore need to look at the other options available to us.

We could contemplate continuing to pollute, but adapting to a rapidly destabilising climate. This would involve defending or relocating many of our great coastal cities at the same time as we pay the ever-rising costs of climatic extremes. It would mean enormous changes to our cities and agriculture, and the inevitable degradation of biodiversity and human health. Adapting to a constantly changing set of conditions in a world of out-of-control climate change is a daunting prospect. Deterred by the immense difficulties of reducing emissions, or adapting to the consequences of the ongoing pollution, some contemplate geoengineering our way out of trouble with schemes to put sulphur into the stratosphere, or pump liquid CO_2 into the ocean depths. I myself have contemplated such things in the past. But I have since decided that using a poison to fight a poison is not a useful method, nor one that meets widespread public acceptance.

As I researched this book the question I kept coming back to is this: is there another way forward—a 'third way' as I've come to think of it, in addition to just adapting or geoengineering? I believe that there is, and that it's fundamental to our future. The 'third way' is a new concept, encompassing proposals and experiments that shed light on how Earth's natural system for maintaining the carbon balance might be stimulated to draw CO_2 out of the air and sea at a faster rate than occurs presently, and how we might store the recovered CO_2 safely.

Some third-way technologies and methods have been previously considered geoengineering, but they are qualitatively different. They do not seek to fight one poison (excess carbon) with another (for example, sulphur). Instead they look to restore or learn from processes that are as old as life itself. The third way is in large part about creating our future out of thin air. That might sound fanciful, but in fact it is the only way that complex life has ever prospered—by building itself out of CO_2 drawn from the atmosphere.

It is vitally important that we understand the difference between third-way technologies and measures that reduce the burning of fossil fuels. One big difference lies in the security of the carbon store. The carbon in fossil fuels has been safely buried for tens if not hundreds of millions of years, and would stay safely locked in Earth's crust if we did not dig it up. But the carbon stored in soils, trees and other vegetation is part of the living carbon cycle. Oftentimes it will remain stored only as long as policies protecting it exist.

Previously, third-way technologies and measures that reduce fossil-fuel use have been conflated in international negotiations, and the result has been disastrous. For example, under the Kyoto Protocol, Australia negotiated to stop land clearing, but did nothing to curb the burning of fossil fuels. As a result, its emissions from fossil fuels rose 30 per cent between the Kyoto baseline year (1990) and the end of the Kyoto commitment period (2012), leaving Australia the largest emitter per capita in the world.[7]

The current Australian policy of 'Direct Action' looks set to lead to a similar outcome. The federal government is spending A$2.55 billion in a series of reverse auctions (where the lowest bids are accepted) to reduce emissions. The first auction, which took place on 23 April 2015, spent much of the A$660 million available on purchasing emission reductions from farmers wanting to pursue third-way technologies.[8] While this will yield valuable experience, it is no substitute for cutting emissions from fossil fuels hard and fast. Instead of being regarded as equivalent to cuts in emissions from fossil fuels, third-way technologies should be seen as a potentially valuable complementary series of options whose full potential may not be realised for decades.

Is it possible for planetary processes to remove greenhouse gases from the air at a scale, rate, permanence and cost that will make a difference to our future climate? The answer, I found, is an emphatic 'yes'. Third-way methods and technologies may, in coming decades, allow us to draw down gigatonnes of CO_2 per annum and safely store it. The third way is not some utopian dream. We may well fail to embark upon it, or we may progress too slowly along its intricate pathways, but it is a direction that it is possible for us to take. Which is why I argue that we must start preparing the ground now, even as we undertake the gargantuan effort of cutting emissions. But the third way needs attention in terms of research, experimentation, funding and political agreement right now if it is to provide mature methodologies and technologies in a few decades' time.

Even if by some miracle we reduce emissions fast enough to preserve our climate safety, the third way might prove useful to us. Its various methods might help restore the atmosphere and the seas to a less perturbed state than would otherwise exist. Instead of average temperatures across the US rising by around 1–2°C, we might be able to limit the rise to 1°C or even less. It's even possible that the third way might bring us unanticipated side benefits for sustainability overall.

Before investigating this hugely exciting hope for our future, we need to know how things stand now. How close is the great climate crisis? Can our desire to overcome it drive humanity's next great waves of positive technological economic and social revolution, or will we be plunged into the dystopian collapses and terrors of civilisations past?

1

CLIMATE SCIENCE

CHAPTER ONE

The Weather Makers: Right or Wrong?

Whether we and our politicians know it or not,
Nature is party to all our deals and decisions,
and she has more votes, a longer memory, and
a sterner sense of justice than we do.

WENDELL BERRY

WHEN I wrote *The Weather Makers*, I laid out the state of climate science as it was understood in 2005. The book received much acclaim, but it was also criticised by climate-change sceptics as extremist and alarmist. One critic even wrote his own book in search of any faults or exaggerations. I was pleased that he found only a handful of errors, none of great consequence: no wide-ranging science book can hope for 100 per cent accuracy. Here I wish to be my own critic, for it's only with the passing of time that it has become possible to test whether claims about climate change have come to pass.

Since *The Weather Makers* was published, the Intergovernmental Panel on Climate Change (IPCC) has completed two major summaries, in the form of its fourth and fifth assessment reports, and thousands of scientific publications have added to our understanding of how Earth's climate system responds to carbon pollution. The IPCC does not do its own research. Under the auspices of the UN it provides a report card based on an enormously thorough examination of the published scientific literature. As a result, many details of climate science have been clarified. Not only are the scientific projections of major trends more certain than ever, but today many of us also have firsthand experience of living in a strongly shifted climate. With climate change an experienced reality, and the science verified, the room for climate change denialism keeps shrinking.

Another thing that has changed over the past decade is the accessibility of climate science to the general public. Ten years ago I had to comb the scientific literature and textbooks for explanations of how our gossamer-thin atmosphere interacts with the oceans and land to create components of the climate system. Indeed I found that books written a century earlier by luminaries such as Alfred Russel Wallace were more poetically eloquent on the subject and more informative than many recent works. Today, a great wealth of material is available to anyone curious enough to look, much of it online (including on the excellent website scepticalscience.com). Sadly, the misleading blogs by sceptics have also proliferated, though an increasingly sophisticated public is less easily deceived by them.

Some of the tools that climate scientists use have also changed enormously. Our capacity to model the climate system in time and space, for example, has been transformed. The models used in the 1990s could operate across four orders of magnitude in time and space, while current ones operate across five orders of magnitude. By way of illustration, four orders of magnitude in space extends from a millimetre to 10,000 millimetres (10 metres), while five orders extends from a millimetre to 100,000 millimetres (100 metres). The capacity of the climate models continues to grow at a rate of around an order of magnitude per decade, with each decadal increase involving 10,000 times more calculations for space alone than were required previously. When climate models are able to operate across 14 orders of magnitude—from milliseconds to millennia, and millimetres to thousands of kilometres—they will be able fully to model Earth's climate system.[1]

Despite their vast increase in computational power, the models remain consistent in telling us that our Earth is warming, and will continue to warm in proportion to the volume of fossil fuel we burn. What has changed is the detail they reveal about the things that will unfold. While no climate model can predict the future—simply because the future is impossible to predict—the increasing computational power of the models means that they are becoming ever more useful at explaining how climatic changes are being influenced by humanity.

Studies of past climates are also becoming ever more informative. One that examines over 1000 years of temperature records has shown that climate trends have sometimes differed markedly in the northern and southern hemispheres. One example of hemispheric difference, which the sceptics used to cast doubt on the fact that CO_2 causes warming, concerns the medieval warm period. The new study demonstrates unequivocally that this warm period was restricted to the northern hemisphere.[2] But such is the unprecedented volume of greenhouse gases that humans have released into the atmosphere that the climate system is being overwhelmed, and today warming is occurring in both hemispheres.

The contemporary world is changing fast; few changes have been as profound or disturbing as the increases in extreme weather experienced right across the planet. For that dwindling band who continue to deny anthropogenic climate change, this is the new battleground—albeit one which is becoming ever more difficult for them to defend. When, in late 2013, Australian Prime Minister Tony Abbott and his environment minister Greg Hunt argued that there is no link between the warming trend and extreme bushfires, they were arguing not only against science, but also contrary to common sense.[3]

The link between extreme weather and climate change is a critical area for public understanding, because it's the devastating extremes, rather than a shift in averages, that have the greatest impact. To deny the link also permits people to believe that climate change is something only for future generations to worry about. It is not. Our climate has already changed, and over the last decade we have begun to witness more frequently the consequences of our profligate burning of coal, oil and gas. Very recent advances have allowed scientists to quantify the human impact on individual extreme weather events.[4] Extremes in the weather are therefore a good place to begin looking at what has changed in climate science over the past decade.

The Australian Open Tennis Championships are Melbourne's moment in the sun, and during the fortnight of the competition there's hardly another topic of conversation in the city. When, during the 2014 Open, a heatwave of unprecedented ferocity struck Melbourne, bringing a record-breaking four days in a row of temperatures over 41°C, as well as the city's hottest-ever 24-hour period, the stadium built to host the event turned into a furnace. Despite the long and loud warnings of the climate scientists that extreme heatwaves were all but inevitable, Rod Laver Arena had not been built to cope with the threat, and lives and money were put at risk.

With millions of dollars at stake, the tournament organisers were reluctant to call an end to play. For day after scorching day the players slogged it out in 40°C+ temperatures on the courts. The fans stuck around too, though more than 1000 had to be treated for heat stress. Finally, the health risks to both players and spectators became too much, and the multi-million dollar tournament was suspended.[5]

Australia's growing heatwave crises rarely make global news, but the suspension of the Australian Open made page one in newspapers around the world. The Australian Climate Commission, which I headed until it was closed by the Abbott government in September 2013, had led in warning that heatwaves were becoming more severe in Australia, so I was not surprised when the BBC World Service approached me to explain the relationship between the tennis-cancelling heatwave and climate change.

I told the journalist how the small shift in average temperatures caused by emissions of greenhouse gases was influencing the extreme temperatures across Australia, and indeed the world. The reporter said he'd call me back. When he did, it was to apologise that the interview had been called off. His bosses had told him that unless I could prove that climate change had caused the heatwave they were not interested in the story. I was astonished. In a system as complex as Earth's climate, single factors are rarely the sole cause of anything, so it's usual to talk in terms of influence rather than cause.

But when it comes to the question of climate change's influence on the Australian heatwaves of 2013–14, we do have a definitive answer. In late 2014 Dr Thomas Knutson of the US Geophysical Fluids Dynamics Laboratory at Princeton University, New Jersey, and colleagues published an analysis demonstrating that it is virtually impossible that the extreme heat experienced over Australia in 2013 could have occurred without the influence of human-emitted greenhouse gases.[6] The analysis used a large series of computer models, some of which exclude human influence, while others include it. The Australian heat of 2013 was so extreme that in the 12,000 simulations generated by the models that included only natural factors, in all but one simulation it lay outside the range of probabilities. Moreover, human influence tripled the odds that heatwaves that year would occur as frequently as they did, and doubled the odds that they would be as intense as they were. Our ability to link some kinds of extreme weather to climate change in this way is very new, and is likely to revolutionise our understanding of how we are influencing Earth's climate system.

The average temperature of Earth's lower atmosphere has risen by just under 1°C during the past 200 years. How, you might ask, can such a small average increase have a large effect on extreme weather? There are several aspects that must be considered. One is that, because around 90 per cent of the extra heat captured by greenhouse gases is transferred to the oceans, the oceans are warming dramatically. This alters evaporation, which influences the intensity of rainfall as well as the intensity of cyclones, and indeed the water cycle as a whole. But a second, more important, answer lies in the simple observation that if you plot weather for any location it looks like a bell curve. As you can see from the diagram below, you need only shift the average temperature a little to have a huge effect at the extremes. The graph, incidentally, is a realistic reflection of the extent to which climate change has shifted the average temperature at many locations in the northern hemisphere over the last half century.[7]

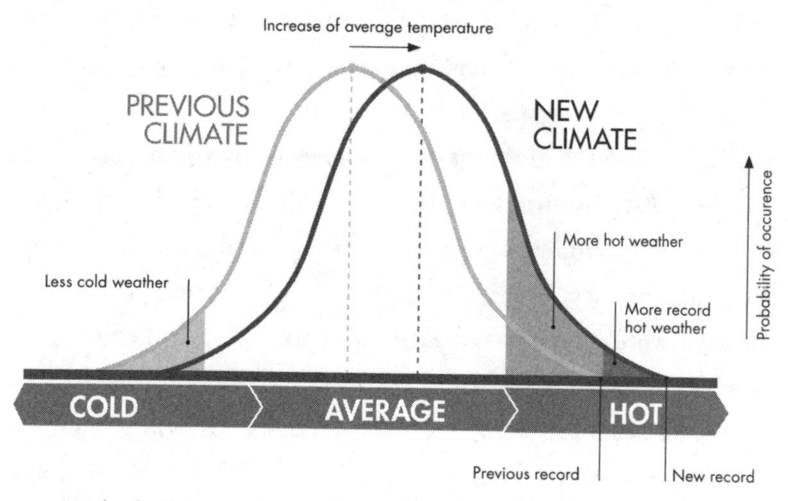

As the bell curve shows, we will still experience some cold days in our warmer climate. But we will get many more hot days, as well as a number of record-breaking hot days. During the summer of 2013, more than 3000 weather records were broken in the US, while 123 such records were broken in Australia (which has far fewer weather stations). In 2014 a further 156 records were broken in Australia.[8] We're seeing the climate change before our eyes.

So, how serious have things got with a global average of less than 1°C of warming? There's no better place to begin investigating than heatwaves, which are defined as prolonged periods of above-average temperature at a location that last days to weeks. They are, increasingly, the most fatal of all changes flowing from the increase in average temperatures.

A well-documented heatwave experienced in Melbourne, Australia, in January 2009 shows in detail how heat affects health. After four days of high night-time as well as daytime temperatures, many people's bodies had become overstressed and unable to shed the excess heat. Mortality records reveal that, on average, around 90 people die annually in Melbourne between January 26 and February 1. But during the heatwave of 2009, 374 'excess deaths' were recorded, the great majority occurring after four days of the extreme heat.[9] Bushfires and hurricanes might gain the headlines, but it's easy to understand why doctors have come to dread what they call 'the silent killer'.

Heatwaves have, of course, always occurred. The dustbowl-era American heatwave of 1936 was the hottest on record until 2012. The great Chicago heatwave of 1995, which killed about 600 people, occurred as greenhouse gas concentrations were beginning to climb, and so may have been influenced by climate change. But it was only with the arrival of the twenty-first century that our shifting climate began to influence heatwaves strongly. Humanity's first intimation of just how great a threat to health heatwaves could become arrived in the summer of 2003. Europe's summer that year was the hottest since records began in 1540.[10] The most severe conditions were felt in France, and by August—the traditional time for summer holidays—parts of the country were sweltering in record heat. With their families at the seaside, many elderly people had little support for coping with the extreme conditions, a situation made worse by the lack of preparedness by authorities.

Because heatwaves had not been considered a major threat, many aged-care facilities lacked air conditioning. Remarkably, though, the residents of nursing homes did better than the more capable elderly living alone at home. With nobody to help them, it was the 'fit elderly' who experienced the worst mortality rate, succumbing to overheating, dehydration and heart and lung failure. In France, nearly 15,000 heat-related deaths resulted in a severe overload of mortuary facilities, and a refrigerated warehouse outside Paris was used as temporary storage. Across Europe, more than 70,000 people died of the heat.[11] At current rates of warming, by mid-century the conditions seen in the 2003 European heatwave are set to become the annual summer average.[12]

In July–August 2010, another heatwave caused an estimated 56,000 deaths, this time in western Russia. The extreme conditions were part of a larger phenomenon, the period April–June 2010 being the warmest on record for land areas in the northern hemisphere.[13]

In Australia, heatwaves are hotter, last longer and come earlier than ever before. The number of hot days (above 35°C) across the country per year has doubled in the last 50 years, and the annual number of heatwaves has doubled in Perth. In Adelaide, heatwaves last an average of two days longer than they did 50 years ago, while in Melbourne the heatwave season is starting 17 days earlier.[14] In 2014, Adelaide experienced 13 days over 40°C (the average is currently two), a number expected for the average summer by 2030.

In the last few years, record-breaking heatwaves have been felt from Shanghai to Texas. In the US in 2011–2012, the number of intensive heatwaves was almost three times the long-term average, with the 2011 Texas heatwave and the 2012 heatwave in the Midwest both breaking temperature records.[15] Akin in their extremity to the Russian heatwave of 2010, they give some intimation of the conditions scientists predict are likely to be felt towards the end of this century in the US if we don't rein in emissions.[16]

At the same time that unprecedented heatwaves are being felt across the planet, recent winters in some densely populated areas of the northern hemisphere have been unusually cold relative to the average of the previous two decades. So what's going on? Some scientists are investigating the possibility that the polar vortex is being weakened by the rapid warming of the poles. The polar vortex is a strong current in the atmosphere that separates the frigid air lying over the Arctic from the warmer air to the south. With a diameter of about 1000 kilometres, it is a persistent and powerful cyclone centred on the North Pole, the strength of which varies seasonally and annually, as it is influenced by many factors.

A weak polar vortex is associated with cold winters in the northern hemisphere. Like a top losing momentum, when a weakened polar vortex encounters intense low-pressure systems, it can't push through, but instead flows around them to the north or south. In effect it develops the wobbles. When it's diverted to the south, as during recent winters over North America, great masses of frigid air push south, while over parts of the Arctic, lobes of warm air penetrate deep over the normally frigid ice cap, melting sea ice and permafrost.

It's easy to see how a decreasing temperature gradient can weaken a cyclone like the polar vortex, which feeds on temperature differences. But many other influences are felt on this complex and variable system, and years of study will be required before firm conclusions can be drawn. For the moment, the default position of many scientists is that the recent cold winters are just part of natural variability. Even as Earth warms, we will continue to experience some cold days and cold seasons.

As heatwaves become hotter, longer and more frequent, there's an inevitable impact on forest fires, or bushfires as they're known in Australia. While the relationship is not as simple as that between rising average temperatures and heatwaves, it is clear enough. Three things are necessary for a forest fire to rage—enough fuel, a source of ignition, and the right weather. Sufficient fuel exists much of the time at many locations, and sources of ignition, from dropped cigarette butts to lightning strikes, are ever present. So, as common sense suggests, the right weather conditions are the key factor in determining how severe a bushfire will become. Few places are as bushfire prone as southeastern Australia, and the most severe and damaging fire in the nation's history provides a clear picture of how climate change is influencing fire risk.

The people of Victoria will not soon forget the conditions they awoke to on 7 February 2009. For a decade, drought had ravaged Australia, and almost no rain had fallen in Victoria in the preceding two months. A heatwave had killed hundreds in the previous week, and temperatures were predicted to peak at 47°C that afternoon. Add to that a scorching southern-hemisphere northerly wind blowing at 100 kilometres per hour, and it was clear to everyone that this would be a day of dread.

Despite all the warnings and preparations, the thousand fires that broke out that day across the state would take 173 lives, almost half of them children. Under the extreme conditions that day, fire began to behave in ways never seen before. The speed of its spread and the intensity of its impact took almost everyone by surprise. The Black Saturday fires, as they became known, were Australia's most deadly bushfires, and among the 10 worst ever recorded worldwide. In their wake, official advice about what to do when bushfires threaten has been drastically rewritten. Before Black Saturday, official advice was often to stay and defend your property. Afterwards, officials now warn residents to evacuate as soon as the announcement is made, which is often long before a fire reaches an area.

It turns out that at almost every step along the way, climate change had some influence on the severity of the Black Saturday bushfires. In Victoria, lightning strike typically accounts for about 25 per cent of all bushfire ignitions but, because bushfires started by lightning often burn in rugged areas and are difficult to put out, such fires account for around half of all land lost to bushfires in the state. And with every degree Celsius of warming experienced, lightning activity increases 5–6 per cent.[17]

The amount of fuel available to a fire can be influenced by climate cycles, which themselves are affected by climate change. Extreme rainfall can promote plant growth, and a long dry spell can then cure it and make it ready for burning. But by far the greatest influence of a changing climate on bushfires are short-term weather events, such as heatwaves. As one climate scientist put it, 'climate change is increasing the frequency and severity of very hot days and driving up the likelihood of very high fire danger'.[18]

In regions as diverse as Spain, Greece, areas of Africa and parts of North America, increases in the extent and intensity of forest fires have been linked to climate change.[19] In Canada, for example, the area of forest burned has increased in the last four decades as summer temperatures have risen. One study estimates that the boreal forests have not burned as extensively in the last 10,000 years.[20] In 2014 massive wildfires burned through the Northwest Territories, scorching an area six times greater than the 25-year average, yet in line with climate projections for increasing forest fires at high latitudes.[21] Similar increases in forest fires have been seen in Spain, where the frequency of fire has doubled and the area burned annually increased tenfold since the 1970s.[22]

How might forest fires play out in future? In Australia, while the projections regarding future bushfire activity are not detailed, most analyses indicate an increase in the fire danger index as temperatures increase.[23] Another issue that is likely to have a large impact is that, as the fire season lengthens, the opportunities to control fires through hazard-reduction burning are decreasing.[24] This leaves communities under threat of high fuel loads in an extended fire season. Fire crews are already finding the longer fire season a challenge. A few decades back they could work together and stage their efforts to meet the fire challenge as it unfolded sequentially across the continent from north to south. Now, they scramble to meet extreme conditions that may erupt in a number of states simultaneously.

Elsewhere, future trends are clearer. Even modest warming is projected to bring substantial increases in fire activity in the Rocky Mountains. Areas that have experienced fire every 200–300 years might see it every 30 years by 2050.[25] In California, increases of more than 100 per cent are expected in many forested areas, with the extreme fires of May 2014 offering yet another example of how rapidly the predictions of the scientists are being fulfilled.[26] In Alaska and western Canada too, conditions are set to worsen from their already severe state, with the average area burned by fires projected to double by mid-century and increase fivefold by the end of the century.[27] The age of the mega-fire, it seems, is about to arrive, if indeed it is not already here.

All of this puts the public at heightened risk of yet another silent killer. Each year, more than 300,000 people die worldwide from inhaling smoke from forest fires.[28] As the fire risk grows, the threat to life will worsen. In the western US, for example, organic carbon aerosol (a component of smoke) concentrations are expected to increase by 40 per cent by 2050.[29] Smoke from forest fires can threaten the health of people far from the fire front. During July 2002, forest fires in Quebec resulted in an up-to 30-fold increase in airborne fine particle concentrations in Baltimore, a city nearly 1600 kilometres downwind. These fine particles are extremely harmful to human health and affect both indoor and outdoor air quality.[30]

Human health can also be directly impacted by the burning of fossil fuels. In northern China, air pollution from the burning of fossil fuels, principally coal, is causing people to die on average

5.5 years sooner than they otherwise might.[31] It's not just the particles produced by burning coal that are deadly. As Earth warms, more ozone is created at ground level, and ozone, when combined with fumes from burning fossil fuels or from wildfires, creates photochemical smog. Almost anyone living in a city will be familiar with the brown haze that results, and those who are young, asthmatic, old or suffering from lung or heart disease will know firsthand of its health impacts. As Earth warms, the problem is bound to increase, and the health effects will be compounded by other factors, including allergies.[32]

In North America, hayfever sufferers dread the ragweed pollen season. Courtesy of global warming, the season has increased in length by between 11 and 27 days in parts of the US and Canada. And when photochemical smog or other air pollutants are present, exposure to them can spark heightened episodes of allergic reactions and asthma.[33]

The increase in heavy downpours resulting from climate change is also affecting health, through promoting an increase in the growth of mould. This compounds lung diseases, allergies and asthma. Heavy rains can also overwhelm drainage systems, causing exposure to sewerage and toxic chemicals that add to other climate-related health issues.[34]

Diseases that are transmitted by insects are also affected by our changing climate. This is hardly surprising, as insects are exquisitely sensitive to the climate. The blacklegged tick, which carries Lyme disease, requires specific temperatures, rainfall and humidity to survive, and in North America the changing climate is delivering it a windfall. At present the tick's range is restricted to eastern and southeastern US, but is predicted to have spread over most of the eastern half of the US and deep into eastern Canada by 2080. The only good news is that parts of Florida and the Gulf Coast will become too hot for it.[35] Similar trends in the spread of Lyme disease are evident in Europe, while in Australia as many as eight million people may become exposed to dengue fever by 2100 as the mosquito that carries it moves further south.[36]

Just whose health is most at risk from such impacts has been assessed in a recent Australian report.[37] Those living in remote areas, including Indigenous communities, low-income earners, the elderly, children, those who work outdoors, those with existing medical conditions, and tourists (who do not appreciate how extreme local conditions can become) are all at elevated risk.

Perhaps the most surprising health impact of climate change was revealed by a study published in *Nature*. It documented how increasing atmospheric CO_2 was degrading the nutritional value of crops, especially in Asia. Zinc, iron and protein levels are all falling in wheat and rice, and at least two billion people depend on rice for their iron and zinc. The researchers compared the nutritional value of crops grown under varying CO_2 concentrations, and found that at concentrations of around 550 parts per million, the amount of zinc in wheat declines by 9.3 per cent, iron by 5.1 per cent and protein by 6.3 per cent.[38] This occurs because the increased CO_2 allows the plants to grow faster and larger than previously, but because there are no more nutrients and minerals, their concentration is reduced. Broadly similar declines were documented for rice, peas and soybeans. Lead author of the study, Samuel Myers, said that the finding was 'possibly the most significant health threat that has been documented for climate change'.[39] Significantly, the article made page-one news in the *China Daily*. Given the Chinese government's control of the media, the prominence granted to the story should be read as preparation for significant new measures in that country to reduce greenhouse gas emissions.

Finally, to all such impacts must be added the mental illnesses caused by stress. Losing your home to a bushfire supercharged by a warming climate is traumatic, as is being flooded out, or losing your crop. Health workers are noting an uptick in people whose mental health is being affected in many ways due to our changing climate.[40]

How will health impacts play out in future? An excellent response to the question was published in 2014 by global health expert Dr Tony McMichael. He says:

> Climate change over the next few decades will mainly act by exacerbating existing health problems. The greatest impacts will occur—indeed, are occurring—in populations already burdened by climate-sensitive health problems such as child diarrhoea, nutritional stunting and urban heat extremes. Human-driven warming has increased heat-related death and illness in many locations, while changes in temperature and rainfall have altered the distribution of some waterborne infectious diseases and reduced food yields in some food-insecure populations.
>
> These adverse health impacts will widen the existing health gap between regions and between rich and poor. Climate change, unabated, will erode development gains—an issue now of explicit concern to the World Bank and... the Asian Development Bank.[41]

Looking beyond 2050, it becomes more difficult to assess risks to human health. But one thing is clear. The impacts of increasing temperatures upon human health are not linear. In a world in which temperatures are 4°C warmer than the pre-industrial base, we are likely to see health impacts many times more severe than those which will prevail in a world 2°C warmer.[42] As a worst case, our current trajectory could even:

> erode the essential foundations of human population health—food yields, water supplies and the constraints on infectious disease rates, population displacement, conflict and war.[43]

CHAPTER TWO

The Waters of a Warming World

We are having a little problem in Miami Beach, we are getting water in the streets. Where do you suggest we put it?

MIAMI PUBLIC WORKS STAFF TO PROFESSOR HAROLD WANLESS, C. 2008–09[1]

THE build-up of greenhouse gases in the atmosphere is having a big impact on Earth's water cycle. An increase in rainfall intensity has been recorded across many regions of the globe in recent decades. Essentially, rainfall intensity is a measure of how much water falls from the sky during a given period. Increases in the proportion of rain falling as heavy downpours have occurred in all regions of the US, with a 79 per cent increase in the northeastern states.[2] Rainfall intensity is influenced by the rate of evaporation and the amount of water vapour in the atmosphere, as well as other factors. Greenhouse gases are a strong influence on rainfall intensity because they warm the oceans and atmosphere. A warmer ocean evaporates more readily, and a warmer atmosphere can hold more water vapour. When rainfall occurs under these warmer conditions, it is likely to be more intense.

Snow is a form of precipitation too. Today the lower atmosphere (where most of the greenhouse gases are) is warmer than it was previously, as is the sea. When warm, moisture-filled air meets colder air higher in the atmosphere, this can result in colossal snowfalls. Extreme snowfalls have been experienced in recent years in many parts of the globe.

Most drainage infrastructure has been built with historical rainfall intensity in mind. For example, cities in the tropics are equipped with relatively large drains, because tropical downpours can be far more intense than rainfall in temperate regions. So, changes in rainfall intensity can contribute to flooding by overwhelming drainage systems and levees built with less extreme conditions in mind. Storm surges can add to these flooding impacts in coastal cities.

In the summer of 2010, Queensland experienced its wettest December on record. Floods broke river-height records at more than 100 observation stations,[3] and 78 per cent of the state was declared a disaster zone.[4] At the time Queensland was experiencing these extremes, the surface temperature of the ocean surrounding northern Australia was the highest on record.[5] The economic impact of the increased rainfall was unprecedented. More than 300,000 homes and businesses in the Brisbane–Ipswich area lost power, and mining infrastructure was put out of action for many months, forcing companies to declare *force majeure*. Total uninsured costs for the Queensland government alone were about A$5.6 billion. In order to help pay the repair bill, the Australian government instituted a one-off levy. The following year, Australians earning $50,000 or more per year paid up to 1.5 per cent extra income tax. There is some irony in this. At the time the flood levy was being mooted, Australia's opposition Liberal Party was busy whipping up public fury at the new 'carbon tax' (in fact a fixed-price start to a carbon-trading scheme), which the Labor government had introduced to help reduce the impacts of climate change.

The Queensland floods of 2010–12 have been examined using newly developed methods to determine whether human-induced climate change exacerbated their intensity. The research reveals that the La Niña event then being experienced was by far the most important factor, although human-emitted greenhouse gases probably exerted a minor direct influence.[6] La Niña is a phase in a climatic cycle that is marked by cooler than average sea surface temperatures in the central and western Pacific Ocean. It often brings increased rainfall to Australia. A more recent research paper, however, points out that La Niña events are themselves being influenced by human carbon pollution, because they accelerate the rate at which the land is warming relative to the oceans. As a result, La Niñas are projected to increase in frequency from the current average of once in 23 years, to one every 13 years. There is also a greater chance of them occurring hard on the heels of El Niño events, creating extreme swings in climate.[7] So the human influence on floods such as those experienced in Queensland in 2010–12 may be felt more strongly in future.

Despite the increase in rainfall intensity experienced overall, shifting atmospheric circulation patterns are decreasing the amount of rain falling in some regions. Areas affected include Africa's

Sahel, America's west, Australia's south, and even flood-prone Queensland. Following the unprecedented 2010–12 floods, in March 2014, 80 per cent of Australia's Sunshine State was drought declared, the greatest percentage ever in its history.[8] While it's true that Australia has always been a land of drought and flooding rains, climate change is influencing these events so that they become more extreme.

Other recent droughts with a documented climate-change impact include those that occurred in Texas and Oklahoma in 2011. Both states experienced their hottest summer since record keeping began in 1895, with many locations experiencing more than 100 days over 37.8°C. Rates of water loss, due in part to increased evaporation, were double the long-term average. The intense heat, combined with drought-depleted water resources, contributed to more than US$10 billion in direct losses in agriculture alone.[9]

When I wrote *The Weather Makers*, a convincing example of climate change's impact on the hydrological cycle was the decrease in stream-flow into Perth's dams up to 2002. A large decline in rainfall in southwestern Australia had most likely been caused by CO_2 pollution and the ozone hole, and rising temperatures were robbing the soil of moisture. By 2001 this had led to a dire water shortage, forcing the Western Australian Water Corporation to plan to build a desalinisation plant. Completed in 2006 in Kwinana, a Perth suburb, the plant now supplies the city with 45 gigalitres (45 billion litres) of water per year. But the situation continued to deteriorate and a second plant, with an annual capacity of 100 gigalitres, was planned. Stage one is now complete, and stage two is under construction.

At the time these huge infrastructure investments were planned, there was a possibility that the scientists were wrong, and that the rain might return. The latest stream-flow figures for Perth's dams, however, show how wise the investments were. In 2012, the dams received just two gigalitres of water, as against the average for the seven years following 2005 of around 70 gigalitres per year. Between 1911 and 1974 those same dams had received an average of 338 gigalitres per year.[10] As of mid-2014, almost half of Perth's water requirement is supplied by desalinisation.[11]

I was in Perth in 2011, when the hysteria about Australia's carbon tax was at its height. The climate-change deniers were hard at work using the 2010 Queensland floods as 'proof' that the climate scientists were wrong in warning about water shortages. Much of the attack was aimed at me personally, so I was grateful when the previous head of the state's Water Corporation thanked me, publicly, for helping raise the alarm about water shortages in Perth. Without such warnings, he said, the investments required for the desalinisation plants might not have been forthcoming, and Perth might have faced a full-blown water crisis.

Other changes in water availability predicted by climate scientists in 2005 have since been validated. In California, global warming is melting the snowpack earlier, robbing agriculture of water when it's needed. In May 2014, authorities announced that the second smallest snowpack in the state's recorded history had formed the previous year.[12]

In fact, as a new study shows, some parts of California's mountains lifted as much as 15 millimetres during 2013–14 because the massive amount of lost snow is no longer weighing down the land, allowing the unburdened mountains to rise, a bit like an uncoiled spring.[13] The snow losses saw the California Department of Water Resources set the 2014 water allocation at a mere 5 per cent of that requested by agricultural and other public water agencies—the smallest allocation in the department's 54-year history.[14]

As weather historian Chris Burt wrote of the evolving situation:

> The next six months are going to be a severe test of the state's ability to manage...We will probably see (and already are) clashes between agricultural concerns and urban consumers. The specter of a horrific fire season also looms over all this. Since the drought of 1975–77 the population of California has almost doubled (from about 20 million to 38 million). The consumption of water resources by the agricultural industry has also dramatically increased.[15]

Recent research suggests that this may be just the beginning of a far larger water problem for the region. A study using data from tree rings (which reveal changes in rainfall going back 1000 years or more) combined with sophisticated computer models, indicates that California, along with much of the Great Plains and American Southwest, is at high risk of 'mega droughts' in coming decades.[16] The tree rings reveal that droughts lasting 20 years or more—far more extreme than anything experienced since European settlement—struck the region 800–900 years ago. The computer models meanwhile indicate that conditions are right for even more severe droughts—lasting up to 50 years—to strike the region later this century. Cornell University's Professor Tony Ault, co-author of the study, said of the findings: 'In both the Southwest and Central

Plains, we're talking about levels of risk of 80 per cent of a 35-year-long drought by the end of the century, if climate change goes unmitigated.'[17]

When I wrote *The Weather Makers*, Africa's Sahel region was in the grip of severe drought. A study by Australia's scientific research organisation, CSIRO, had proposed that an important factor in the Sahel drought was 'global dimming' caused by particulate pollution produced by factories and motor vehicles in Europe and elsewhere.[18] The unfolding story has become more complicated. Changing sea-surface temperatures have now been implicated, via a natural cycle known as the Atlantic Multidecadal Oscillation (AMO).[19] The AMO describes cyclical changes in the surface temperature of the Atlantic Ocean and is associated with changes in thermohaline circulation (of which the Gulf Stream is part). The AMO's warm phase is expected to peak in about 2020, which may bring temporary relief to the region's water shortage.[20] Another factor affecting the drought, however, is record-breaking high temperatures. In 2010, extreme temperature records were broken in Chad, Niger and Sudan, the new records all being between 47.1°C and 49.6°C.

One area where the global climate outlook has improved somewhat concerns hurricanes. By 2005 it was evident that the number of hurricanes in the North Atlantic was increasing. They are also becoming more intense and extending into regions, such as the south Atlantic, where they had not previously occurred. In the light of these trends, some scientists anticipated that the number of hurricanes worldwide would increase. But hurricanes are relatively rare weather events, making it difficult to get statistically significant samples over time and space, and other researchers felt that the data were not adequate to decide one way or the other.

Over the past decade the situation has become clearer. While the number of hurricanes in the North Atlantic has increased, the total number of hurricanes occurring worldwide has not, nor is it expected to. Hurricanes have, however, increased in severity, and they are occurring further from the equator. As the IPCC recently said in its fifth assessment report:

> It is virtually certain that there has been an increase in the frequency and intensity of the strongest tropical cyclones in the North Atlantic since the 1970s. In the future, it is likely that the frequency of tropical cyclones globally will either decrease or remain unchanged, but there will be a likely increase in global mean tropical cyclone precipitation rates and maximum wind speed.[21]

The reason that the number of cyclones might decrease in future results from the relative influence of several consequences of global warming that work against each other when it comes to cyclone formation. The most influential is the way that the warming of the oceans and the land alters wind direction and speed. This can create wind shear, which tears into the embryonic cyclone vortex and dissipates it. Other factors affect the temperature difference between Earth's surface and the upper troposphere (the top of the lowest layer of the atmosphere). The greater the difference, the more likely it is that a cyclone will form. The upper troposphere is warming faster than Earth's surface because the greater evaporation caused by the warming is carrying increased amounts of water vapour high into the troposphere. When the water precipitates out in the upper troposphere, it releases latent heat energy. But the warming is also causing the troposphere to expand, providing a longer distance over which air can cool as it rises. At present, the latent heat energy provided by the extra water vapour more than offsets the cooling caused by the additional distance the air travels. The smaller temperature difference between the upper troposphere and Earth's surface that results is likely to see fewer cyclones form. But other factors cause regional variations.

The US National Assessment Report summarised the situation for North America thus:

> By late this century, models, on average, project an increase in the number of the strongest (Category 4 and 5) hurricanes. Models also project greater rainfall rates in hurricanes in a warmer climate, with increases of about 20 per cent averaged near the center of hurricanes.[22]

While a Category 1 hurricane may strip the leaves from trees, a Category 5 will uproot them. The extra rainfall, combined with the greater battering caused by rising sea levels, is likely to make future hurricanes a more severe threat to coastal cities. The kind of damage we'll see more of was illustrated by Hurricane Sandy, which devastated the West Indies and the east coast of the US. With a diameter of 1770 kilometres, it was the largest Atlantic hurricane on record and the most destructive hurricane of the 2012 season. But, most significantly, with cumulative damages reaching US$68 billion, it was also the second most costly hurricane (after Katrina) in US history.

When it hit New Jersey, Sandy was only Category 1. Yet it drove a water surge of almost 4.3 metres at Battery Park, flooding large parts of Lower Manhattan. Subway flooding and prolonged interruption to the gas supply were just two of the impacts that left the city reeling. The extent of future damage to coastal infrastructure by hurricanes will be strongly influenced by the extent of sea-level rise, and there things are looking grim.

With just under 1°C of warming experienced to date, the world's oceans are rising at the rate of 3.2 millimetres per annum. Sea-level rise has two components: ice melts, adding to the sea volume, and the expansion of water as it warms. The oceans are absorbing about 90 per cent of all the heat captured by the extra greenhouse gases in the atmosphere. Heat transfer into the ocean is reasonably well understood, and scientists are confident in their predictions of how the thermal expansion it creates will influence sea levels. By the end of the century the oceans will rise by between 11 and 43 centimetres due to heat transfer alone.

The other component—the water added by melting ice—is far harder to predict, as it depends on how the ice caps respond to the warming. This area of research is developing rapidly, with new findings being announced every year. A decade ago it was clear that the Arctic was in trouble. It was warming twice as fast as the planetary average, and the Arctic ice cap, which is sea-ice and therefore doesn't affect the level of the sea as it melts, was vanishing fast. The Arctic ice does, however, help insulate the Greenland ice cap, and its melt waters are likely to add significantly to the level of the oceans. Year-on-year variation of the extent of Arctic ice is considerable, but a long-term trend of dramatic decline is now clear.[23]

As I feared in 2005, the rate of melt of the Arctic ice has proven to be of the runaway type. In the last decade the rate of ice loss over the Arctic has exceeded even the worst-case scenario of the climate models. We now expect to see the Arctic's first ice-free summer in over a million years sometime between 2040 and 2050. An ice-free summer in the Arctic is defined as one with less than a million square kilometres of ice, as the ice around the edges of the Canadian islands melts less readily.

Until very recently, great uncertainty surrounded the state of the Antarctic ice cap. Scientists were having a hard time assessing the contribution of its ice melt to sea-level rise. Accordingly, the IPCC declined to include figures for total sea-level rise in the main body of its Third Assessment Report published in 2001.[24] Climate-change denialists then deceptively used the raw figures to argue that the IPCC had revised the rate of future sea-level rise downwards. They also argued that there was no need for concern about sea-level rise because the extent of sea ice surrounding Antarctica was growing. Eighty per cent of Antarctic sea-ice melts away each year, then grows again.[25] During 2008–11 its minimum extent was small (2.5–3.2 million square kilometres), but between 2012 and 2014 it averaged 3.6–3.9 million square kilometres. Its maximum extent is also variable, and during the winter of 2014 it reached 19.8 million square kilometres—its largest extent recorded since satellite monitoring began in 1979.

Nobody knows why the extent of Antarctic sea ice was so great in 2014, but among the factors that might be affecting it are the ozone hole (which influences wind that can push ice northwards, thereby increasing its extent), the warming of the atmosphere (enabling it to hold more water vapour, which then falls as snow, causing fresh water, which freezes more easily, to collect on the sea surface) and changes in ocean current circulation, which can bring cold seawater to parts of the surface. A recent study even suggests that changes in wave action might also be a factor.[26] As with so much about the Antarctic, much research needs to be done before we have a clear picture about what is causing the variability.[27]

The story of the Antarctic land ice is much clearer, thanks to a series of discoveries published during the first half of 2014. In *The Weather Makers*, I said: 'The increased precipitation occurring at the poles is expected to bring more snow to the high Antarctic ice cap, which might compensate for some of the ice being lost at the continent's margins.' Sadly, that compensation has proved illusory: it is now established beyond doubt that losses of land ice are occurring across all the major regions of Antarctica.

This new understanding comes courtesy of the European Space Agency's Cryosat 2 spacecraft, which was launched in 2010 specifically to measure the thickness of polar ice. It uses a special kind of radar, known as a synthetic aperture interferometric radar altimeter, to chart the surface shape of ice sheets, and it concentrates particularly on their margins. The new assessment incorporates three years of data, 2010–13, and updates a synthesis of observations made by other satellites over the period 2005–10, providing the first accurate, continent-wide assessment of Antarctic ice.

The ice losses Cryosat 2 detected over Antarctica are gargantuan—in the order of 160 billion tonnes a year—the equivalent of two centimetres of snow off the entire surface of Antarctica. That's twice as large a loss per year as when the continent was last surveyed in 2005, and sufficient to add almost half a millimetre of global sea-level rise annually.[28]

Most of the ice loss is occurring in West Antarctica, where the ice sheets have long been known to be less stable than those in East Antarctica. A decade ago I expressed fears that the West Antarctic Ice Sheet (WAIS) may destabilise and melt into the sea. Were this to occur, it would add about 4.8 metres to the level of the oceans.

The Pine Island Glacier (PIG) and the associated Thwaites Glacier are important elements of the WAIS. Between them, they hold enough water (as ice) to contribute a metre of sea-level rise to the world's oceans. In 2014 a study revealed that the PIG is dead on it its feet.[29] Due to a warming ocean, PIG is melting away from below, and nothing can save it. It's important to understand that this finding does not result from computer modelling, but straight mechanics. The undersurface of the ice has melted to the point where the bedrock slopes back towards the glacier's head. That means that the remaining ice will detach from the bedrock and slide into the sea, melting as it goes.

NASA recently analysed 40 years of observations of six big glaciers (including PIG and Thwaites) that drain into Amundsen Bay. It concluded that nothing now can stop them all melting away.[30] Just how long it will take is a big unknown. If measured in centuries it would be a blessing, as the loss of the ice will add around 1.2 metres to the ocean level. At one estimate, during the next 20 years, a 20 per cent melt of ice is likely to add 3.5–10 millimetres to sea levels.[31] As NASA's Eric Rignot says, IPCC estimates of sea-level rise do not take these new data into account. As a result, we should expect sea-level rise to be at the high end of the range estimates for this century.[32] The IPCC's current anticipated range for sea-level rise, incidentally, is between 0.4 of a metre and one metre by 2100.

Another new study lends credibility to the idea that Antarctica's coastal ice may melt more quickly than expected, with consequences for sea level. The researchers discovered that human pollution is causing a strengthening of westerly winds, and shifting them polewards as they circle Antarctica. The trend has been observed since the 1950s and is producing:

> an intense warming of subsurface coastal waters that exceeds 2°C at 200–700 m depth...This analysis shows that anthropogenically induced wind changes can dramatically increase the temperature of ocean water at ice sheet grounding lines and at the base of floating ice shelves around Antarctica, with potentially significant ramifications for global sea-level rise.[33]

The link between winds and warming occurs through a phenomenon known as Elkman pumping, whereby the wind creates a surface stress on the ocean, which propagates a spiral of currents in the water below. This means that the westerly winds circling Antarctica can, at a certain depth, generate a southwards flowing current which draws in warmer water. These findings are so new that their full implications are as yet unexplored. But we would be wise to assume in the face of these cumulative findings that sea-level rise will be a greater problem—perhaps a far greater problem—and that it will occur sooner than we imagined.

CHAPTER THREE

Ominously Acidic Oceans

If we continue emitting CO_2 at the same rate, by 2100 ocean acidity will increase by about 150 per cent, a rate that has not been experienced for at least 400,000 years.

UK OCEAN ACIDIFICATION RESEARCH PROGRAMME

I WROTE in *The Weather Makers* that the increasing acidity of the oceans might not affect shelled marine creatures for several hundred years. Today, ocean acidification is widely understood to be one of the most insidious aspects of CO_2 pollution. Almost all of the most important scientific findings on this issue have been published in the past seven years. The problem arises from the rate of change. If atmospheric CO_2 levels rise slowly, the oceans can compensate by absorbing the extra carbonic acid created by the CO_2. But the current rate of CO_2 increase is the fastest in Earth's entire recorded history.[1] That means that natural processes that maintain the acid balance of the ocean are overwhelmed, and the excess CO_2 increases the ocean's acidity.

The chemistry of ocean acidification is so complex that even scientists in the field have difficulty explaining the precise mechanisms. But in basic terms some of the CO_2 absorbed into the oceans reacts with water and carbonate ions to form bicarbonate ions (an intermediate stage in the formation of carbonic acid). Carbonate ions, which are used in this process, are necessary for creatures that build shells, and if there are fewer of them, the job of laying down a shell becomes harder and takes more energy. Because cold water can absorb more gas than warm water, acidification is being felt first in waters near the poles.[2]

The first indication that high atmospheric levels of CO_2 could affect marine life came from an ambitious experiment known as Biosphere 2. Funded by billionaire Ed Bass and opened in 1991, it took place in a 1.2-hectare ziggurat of glass built in Arizona. With its own air, water and nutrient supply, and miniaturised ecosystems of all sorts, it was completely cut off from the outside world. Its purpose was to help scientists understand what would be required to establish a colony on Mars. Ultimately things went badly wrong. Oxygen levels plummeted, and CO_2 soared to over 10 times the concentration in the air outside.

When the project collapsed in 1995, Columbia University took over and began to analyse how and why Biosphere's miniaturised ecosystems had failed to keep conditions habitable.[3] Chris Langdon was given the task of working out what happened to Biosphere's 'ocean'—an Olympic swimming pool–sized tank filled with corals, fish and other sea creatures. Most had died, but there were a few survivors.

Langdon's experiment was basic. He varied the acidity of water that corals grew in, and weighed them regularly. The less acid the water, the faster they grew. When he published his results in 2000, he was able to show that 'coral reef organisms do not seem to be able to acclimate to changing saturation state… [of CO_2]'.[4] Langdon did not realise it, but he was building on the work of a young student of the American zoological pioneer Alex Agassiz. Alfred Mayor had conducted research nearly 80 years earlier on Australia's Great Barrier Reef, and had discovered that corals were highly sensitive to changes in their environment, speculating in 1918 that 'those forms which are sensitive to high temperature are correspondingly affected [by] the influence of CO_2'.[5]

An understanding of what a more acid ocean will be like doesn't depend on what happened in Biosphere 2 or on complex computer models. It can be seen in the places where CO_2 seeps into the water from volcanic vents. These vents create an acidity gradient, which ranges from the acid levels typical of the oceans today, to levels that might be widespread, if we don't act to reduce emissions, centuries in the future.

The best-studied vents are those around Castello Aragonese in the Tyrrhenian Sea, 40 kilometres west of Naples. Far from the vents, the rich patchiness of marine life typical of the Mediterranean prevails, but the closer one gets to the vents, the more uniform, species poor and dominated by algae the life of the sea floor gets. In the immediate vicinity of the vents, due to the highly acid waters, entire classes of organisms are missing, including sea urchins and snails that eat algae. And in the absence of these grazers, algae, fed with excess CO_2, proliferate.

The extent to which different sea creatures are affected by acidification varies widely. Sea urchins are killed by moderate levels, while barnacles are tough enough to hold out even at high levels. Most creatures lie between these extremes, but shellfish, such as oysters, and other economically valuable fish species are severely affected, either dying, or growing far more slowly than usual, as acidity increases.[6]

So how bad is the problem? Two hundred years ago the pH of the ocean's surface was 8.2. Today it is 8.1. It seems like a tiny change, but acidity is measured on a logarithmic scale: that 0.1 difference in pH means that the oceans are 30 per cent more acidic today than they were before the Industrial Revolution. How that translates into trouble for marine creatures is a highly complex and active field of research.

Along the northwest Pacific coast of North America, naturally acidic deep-ocean waters come close to the surface. As early as 2005 scientists had predicted that ocean acidification would manifest itself here early on. In 2007–08, a mysterious pestilence struck the two major spawning facilities that supply oyster spat to the region's oyster farmers. Whole crops of larval oysters were dying overnight, and initially no cause could be found. It was only when scientists from the University of California were called in that things became clear.

With naturally acid waters close to the surface, the pH balance is easily upset. As CO_2 from the atmosphere acidified the surface layer of the ocean, only a small 'sandwich' of less acidic water remained between it and the acidic deeper water. Overnight, the kelp and other plants that abound in the region's waters transpired (effectively breathe out) CO_2, acidifying this sandwiched layer, so that the entire water column was now acidic. When, in the early hours of the morning, the hatcheries brought ocean water in through their pipes, the acid quickly killed the oyster larvae.

There was, incidentally, a happy outcome—a form of adaptation—to this problem. The scientists advised the hatchery managers to leave their intake pipes closed until later in the day. By then, the kelp are photosynthesising and so drawing CO_2 out of the water, making it less acid. The change delivered a great result for the industry. But unfortunately it did nothing for the wild-living marine creatures of the Pacific northwest. They must cope unaided with the acidic conditions created by our CO_2 pollution.

It's only been about a decade since the problem of ocean acidification was first identified, and five years since methods for studying its impacts on marine environments were standardised. Very early studies (conducted in 2007) on the impact of acidification on coral reefs indicated that a doubling of atmospheric CO_2 above the pre-industrial average (likely to occur in the next few decades) would reduce by 40 per cent the ability of corals to form their skeletons.[7] Recently it's been shown that some kinds of corals are less sensitive than others, and that the net average impact of a doubling of CO_2 might be a decline of 14 per cent or less.[8]

A new discovery is that acidity is likely to affect every stage of coral's life cycle, yet almost no studies have been published on the effect of acidity on coral larvae. Moreover, it turns out that you can't really study acidification in isolation. Its impacts may be lesser or greater depending on sunlight, temperature and the presence of oxygen. Despite all the uncertainty in this rapidly developing field, one thing is clear. More acid is never better for marine animals with skeletons and shells. That's because finding the ever-rarer carbonate ions in the more acid water drives up the energy costs of laying down calcium carbonate. So from prawns to oysters, we're likely to taste and feel the cost of those old coal-fired power plants and inefficient cars on our dinner plates, and in our food bills.

One unexpected ramification of acid oceans was recently discovered by British researchers investigating the lugworm, a species commonly used as bait. It turns out that more acidic conditions increase the lugworm's uptake of copper. This toxic metal not only inflicts DNA damage, but also affects the lugworm's sperm, inhibiting reproduction.[9] It is unclear as yet just how many marine creatures are affected by copper poisoning brought on by acid oceans, or indeed how the uptake of other metals such as mercury and lead might be affected, but scientists have raised concerns that many metals may affect marine life as the ocean acidifies.[10] A study (unpublished as of February 2015) of sea urchins shows that they too are affected by the copper poisoning in acidic water.[11] Given the potential impacts, these studies must be taken as a red-flag warning that ocean acidification threatens the very foundations of the ocean ecosystem, and thus our food supply.

Recent experiments in China have shown that in certain circumstances seaweed can be used to restore seawater to a less acidic state. The waters of the Yellow Sea off Lidao Town in north-eastern China are famous for their yields of the edible seaweed *Laminaria japonica*. Seaweed farms there cover about 500 square kilometres of ocean surface, and they yield around 400,000 tonnes of product annually.[12] Because the fast-growing seaweed takes in CO_2 through photosynthesis as it grows, and is removed from the ocean at harvest, it is an excellent means of removing the acidifying

CO_2 from the water. Indeed the seaweed farms of Lidao more than reverse the local acidification threat, providing a safe and nurturing environment for shelled creatures. In other parts of China, where seaweed is farmed in conjunction with scallops, the buffering of the seawater provided by the seaweed provides an environment in which the scallops thrive.[13]

Seaweed is hugely productive, outstripping the fastest-growing land-based crops many times over in its rate of growth and CO_2 absorption. Globally, the potential scale of seaweed farming is 600 times greater than any other method of cultivating algae.[14] Seaweed is finding many uses beyond food, from medicine to fuels, and it may be that seaweed farms will offer refuges for marine species under threat from increasing acidification. One study asserted that seaweed farming could produce 12 gigatonnes per year of biomethane, while storing 19 gigatonnes of CO_2 per year directly from biogas production, plus up to 34 gigatonnes per year from carbon capture of the biomethane combustion exhaust gas. All of this could come from seaweed 'forests' covering an area equal to 9 per cent of the world's ocean surface. This would produce enough biomethane to replace all of today's needs in fossil fuel energy, while removing 53 gigatonnes of CO_2 per year from the atmosphere, thus more than offsetting all human CO_2 emissions. A side benefit would be an increase in sustainable fish production, providing 200 kilograms per capita per year of fish for a population of 10 billion.[15]

While the potential of seaweed to reduce both acidification and global warming is huge, we are very far from being able to exploit it. All marine plants, from seagrass to giant kelp and even humble seaweed, help reduce acidification. Yet never have they been more threatened. All round the world coastal infill, dredging and pollution are threatening the ocean meadows. Unless we address these problems, all the aquaculture presently deployed could be insufficient to compensate for their loss. In fact, given current trends, I have an ominous feeling that the acidifying oceans might just turn out to be the greatest threat of all.

CHAPTER FOUR

How Are the Animals Doing?

We'll lose more species of plants and animals between 2000 and 2065 than we've lost in the last 65 million years.

PAUL WATSON

MORE than 20 years ago, biologists Richard Leakey and Roger Lewin announced that the twenty-first century would be the age of the sixth great extinction.[1] A few years later, climate scientists were warning that, on the current trajectory, two or three out of every five living species may become extinct as a result of global warming. Recent studies have come up with different estimates, making this an area of active scientific debate. But there is no doubt that the current rate of extinction is far higher than the average for Earth, and that many species are imperilled by climate change and other factors. So where is the danger looming closest?

The world's greatest coral reef—Australia's Great Barrier Reef—stretches about 2300 kilometres along the continent's north-east coast, encompassing an area roughly half the size of Texas. Those who have dived its pristine reaches know firsthand that it is one of Earth's natural wonders—a place of exceptional beauty and diversity.

Despite the scientific warnings, a decade ago I found it difficult to believe—even to comprehend—that the world's coral reefs might be on the brink of collapse. The Great Barrier Reef had endured for millions of years, and was protected under law, so I reasoned that surely it was well positioned to survive. I can see now how naïve I was in believing that it might withstand the multipronged onslaught our species has unleashed on it.

In the 1960s and 1970s mining the reef for fertilisers and drilling it for oil were proposed. These threats led to the reef's legal protection. Yet it's now clear that, despite a ban on drilling, fossil fuels have been conducting a lethal stealth attack on the reef. The first intimations came in the 1970s, when areas of coral turned white, then died. Coral bleaching—as the phenomenon is known—occurs when underwater heatwaves stress the coral polyps, causing them to eject the algae living in their tissues, and so turn white. Without algal partners the coral polyps cannot grow the bony skeleton that forms the reef. Indeed they cannot even properly feed themselves. Over a period of weeks the coral polyps slowly starve, then die. When added to the threat of ocean acidity, the attack is devastating. So it is that heat and acid, from atmospheric CO_2 caused by burning fossil fuels, are killing the reef. And it is happening fast.

The reef's current champion is Dr Charlie Veron of the Australian Institute of Marine Science. He says he saw his first bleached coral—a ten-centimetre square patch—off Palm Island in the early 1980s. Now he says, it's 'horrible to see—corals that are four, five, six hundred years old—die' from the heat.[2] For the reef, Veron says, catastrophic global warming has already arrived.

A century ago, a pioneer of coral reef science William Saville-Kent made a photographic record that provides a poignant historic benchmark of the Great Barrier Reef's decline. He was always careful to keep some landmark in the background, so the locations of his photographs can still be traced. We see that a delightful coral garden of a century ago is today a scene of utter devastation. In 2012 a study revealed that half of the Great Barrier Reef has already been killed.[3] Not all the damage has been done by acid and heat, yet, as the years go by, these emerge as the overwhelming culprits.

At the rate at which we are currently burning fossil fuels, the world will be around 4°C warmer by 2100 than it was in 1800. A decade ago I dared to hope that the reef might survive. Certain strains of zooxanthellae (the algae that live in coral polyps) that can tolerate higher temperatures would spread, I thought, helping the coral to live. Or the reef might migrate southwards. A recent study has dashed my hopes. It shows that, if the Great Barrier Reef were to keep pace with a 4°C rise in temperature, its complex ecosystems would need to migrate southwards at the rate of 40 kilometres per year. Yet corals seem unable to migrate at rates greater than 10 kilometres per year. If we do nothing, global warming will simply outpace the reef.[4]

Even if we slow the rate of change, the damage will be monumental. Scientists foresee that 'the majority of existing coral reef ecosystems are likely to disappear if the average global temperature rises much more than about 1.5°C above the pre-industrial level.[5] Through inaction over the past decade we've already assured that global temperatures will rise more than 1.5°C, and even the Paris meeting is aiming only to limit warming to 2°C. It fills me with despair to admit it, but my beloved Great Barrier Reef is doomed. My head tells me what my heart won't. If we exert ourselves to the utmost to reduce CO_2 pollution, the reef may still be able slowly to grow, and even to remain beautiful in patches. But, as an extensive ecosystem, it must be counted among the living dead.

The growing scientific certainty that Australia's Great Barrier Reef, along with many other coral reefs worldwide, will be destroyed by climate change this century is just one example of a natural world imperilled by growing heat and acid. The polar regions are warming faster than anywhere else on earth, and their biodiversity is being strongly affected. For almost everyone who is not an Arctic ecologist, the story of its biodiversity is encapsulated in the fate of the polar bear. Indeed, the creature has become the poster child of climate change.

There was a time when polar bears seemed to feature on the cover of every publication dealing with our climate. Unsurprisingly, it wasn't long before the sceptics began to argue that the story was a beat-up. The great white bear was doing fine, they asserted, and the scientists were simply out to feather their own nests with grant money. Inconveniently for the sceptics, anecdotal but eye-catching stories about the bears, which suggested they were in trouble, kept grabbing headlines. Bears were found drowned at sea, apparently while trying to swim hundreds of kilometres from land to the retreating sea ice edge. And they began hanging around settlements to scavenge, leading to all sorts of run-ins with people.

Recently, tales of cannibalism, usually of male polar bears killing and eating cubs, have hit the news, along with stories of bears eating goose eggs, or climbing near vertical cliffs to feed on guillemots. Bears have even been seen engaging in long-distance chases of reindeer.[6] For animals that predominantly feed on seals, which they catch by waiting beside their breathing holes in the ice, such behaviours are unusual, to say the least. But they do not constitute hard evidence that the bears are starving *en masse*.

News of the 'pizzly'—a polar bear hybrid with a brown bear—is another phenomenon that raised eyebrows. It suggests that the great white bears are travelling far south, or that brown bears are journeying north. There is a concern that the polar bear might eventually hybridise itself out of existence.

To scientists, such anecdotes and rare cases are interesting, but are side issues when it comes to estimating the creatures' chances of survival. The factor that will decide the fate of the bears, they say, is habitat. And for polar bears that means sea ice of about a year old. If the sea ice doesn't last that long, seals don't breed there, or if they do the seal pups drown. And if the ice is more than three years old, it becomes too thick for the seals to make breathing holes in, and so is unsuitable for them.

In the decade preceding 2012, the Arctic sea ice was shrinking even faster than the computer models had predicted. But in 2013 and 2014, as a result of colder winters, there was some recovery—a 12 per cent increase in ice extent above its all-time low in 2012. It's now clear that the main factor controlling the sea ice is temperature, rather than wind and waves, which some previously thought influential. Cold winters will doubtless recur in the future as part of natural variability, but the trend is clear. Warm conditions are predominating, and as a result the ice is melting fast.

Already some populations of bears in the southern reaches of the species' range are starving and failing to reproduce. The southernmost population—in Hudson Bay—has declined from 1200 to 800 in recent years. This is not without precedent. Thirteen thousand years ago the great white bears thrived in southern Scandinavia. But they disappeared at the end of the ice age, when the ice in the area melted too early in the spring or failed to form at all.

A 2010 report by the polar bear specialist group at the International Union for the Conservation of Nature found that eight of the 19 sub-populations are estimated to be in decline, while one is increasing. The other sub-populations are so poorly studied that their status is not known—which is hardly surprising given the remote habitat and rareness of the creatures.[7] The population that is increasing inhabits Davis Strait, between Greenland and Baffin Island in Canada. There, it's so cold that in times past the sea ice often survived for years, becoming too thick for the seals to use. A warming climate is now providing an increase in one-year-old ice in the area—ideal for seals and therefore polar bears.

So controversial is the fate of the polar bear that the 2010 report attracted immediate criticism.[8] Polar bear expert Mitch Taylor, who is much quoted in the media, does not accept the climate science linking the burning of fossil fuels with Arctic ice melt. He says that the Inuit tell him that polar bears have never been so abundant. But this may be because a decrease in hunting is helping some populations, or starvation might be bringing bears closer to settlements where they are more readily noticed. As happens with many sceptics, his comments are often given disproportionate weight by the media in the pursuit of 'balanced' reporting.

Taylor's arguments are frequently linked with observations about the difficulty of estimating the size of polar bear populations. Estimates are often vigorously disputed by climate-change deniers, by hunters (who want quotas increased) or by others with their own agendas. It's clear that research on polar bear numbers has become as politicised as climate change itself. Indeed, as the writer Jon Mooallem comments in his book *Wild Ones*, the debate about the future of the polar bear has become exceptionally bitter, and the waters are now so muddied by misinformation that the general public is confused about whether the bears are endangered or not.[9]

But the facts remain indisputable. Polar bears feed on seals, which need one-year-old sea ice that survives long enough for them to breed on, and overall the Arctic sea ice is melting fast.[10] The bears of Davis Strait may thrive for some decades yet, but if we keep up the current rate of CO_2 emissions, the warming must eventually threaten them as well. Areas now dominated by three-year-old ice, will give way to thinner ice, which will in turn give way to open sea. The most rigorous study predicting the fate of the polar bears was made in 2007 by a team from the US Geological Survey.[11] Based on the rate of melting of the sea ice, it foresaw that the bears would become extinct everywhere, except perhaps in the Arctic Archipelago, towards the end of the twenty-first century.

At the southern end of the Earth, penguins take the place of polar bears as climate-change icons, and the penguins inhabiting the western side of the Antarctic Peninsula have experienced some of the most rapid climate change ever recorded. Over the past 50 years, the average temperature in the region has increased 7°C, and the sea-ice season has shortened by about 100 days since 1978.[12]

The Adélie penguins of the Palmer area of the Antarctic Peninsula are the subjects of long-term study, making their fate as a consequence of climate change the most thoroughly understood of any polar species. The Adélies feed on krill, which take seven years to reach sexual maturity, and which depend on ice to feed. With both ice cover and krill on the decline, the Adélie population has plummeted by more than 80 per cent. In 2012 only 2411 breeding pairs remained, down from 15,202 in 1974. Gentoo penguins, which feed on other prey in the open ocean, are now colonising the area.[13] Studies reveal that abrupt, climate-related changes have upset an ancient stability. One of the five Adélie penguin colonies in the area, which had been in existence for at least 500 years, vanished in 2007, probably as a result of increased snowfall due to the warming, which covered their nests and young, as well as a food shortage due to sea ice loss.[14]

Many other ecological changes are being felt on the Antarctic Peninsula; for example, the kinds of organisms found on the sea floor are changing, and whale and seal numbers are shifting. Not all of this is being driven by warming, but it is a very important factor in altering these ecosystems. So profound are these changes that ecologists believe that the biodiversity of the Antarctic Peninsula will never recover to what it was just 50 years ago.

If changes at the poles have unfolded pretty much as the scientists predicted, what about the 'third pole', the icy peaks of Earth's more temperate regions? Earth's alpine regions harbour exceptional biodiversity. From rhododendron bushes that look like moss mounds to birds of paradise, they are home to species found nowhere else on Earth. And, on peaks not sufficiently high to provide a true alpine habitat, upper montane forests that support unique species are vulnerable in a warming world. All the creatures residing in such habitats are effectively stranded on mountaintops. As Earth warms, they have nowhere to go.

In 2003, Steve Williams, a scientist from James Cook University, predicted that the lemuroid ringtail possum, among other mountain-dwelling species inhabiting the world-heritage wet tropics area of far northern Queensland, would be driven to extinction by climate change.[15] The population most immediately vulnerable lives on Mount Lewis in far northern Queensland. It tends to have white fur rather than the normal sombre grey-brown. With their luxuriant, pure white coat, black eyes and long bushy tail, the Mount Lewis lemuroids are astonishingly beautiful. They are susceptible to climate change both because their habitat is restricted, and because they are acutely sensitive to heat stress, being unable to tolerate more than four or five hours of exposure to temperatures above 30°C. The danger lies not in the slowly increasing average temperature, but in extreme events. Like all of Australia, the rainforests of northeast Queensland are now subject to unprecedented heatwaves.

In 2005, a heatwave hit Mount Lewis and the population of lemuroids crashed. The possums were once so abundant that spotlight searches had usually logged one individual per hour of searching, but in the three years after 2005 repeated lengthy searches failed to spot a single individual. But then, in 2009, three lemuroids—all of the brown rather than the white form—were seen on Mount Lewis during a spotlighting survey. And that, I'm grieved to say, was the last recorded sighting.

In researching this book I was perplexed at how difficult it was to find up-to-date information on the Mount Lewis lemuroid. The Department of Environment and Heritage's protection for the species mentions climate change as a 'suggested' threat, while a recent management strategy for the national park that includes Mount Lewis makes only the briefest mention of the possum and no mention at all of climate change. On coming to power in 2012, the conservative Queensland state government sacked or moved almost the entire staff of the Office of Climate Change, and prepared to close most of the state's climate change and renewable energy programs. I have been told by disgruntled staff of the Queensland Parks and Wildlife Service that they were directed not to mention climate change in official documents. It's hard to avoid the feeling that the public is being kept in the dark about the fate of the lemuroid ringtail possum.

Elsewhere in the world the fate of mountain-dwelling species sensitive to heat is clearer. The American pika, a small mammal related to rabbits and hares, has adapted to life in mountainous areas that rarely get above freezing. They can die when exposed to temperatures as mild as 25°C. American pikas occupy alpine habitats in Colorado, Oregon, Washington, Idaho, Montana, Wyoming, Nevada, California and New Mexico, as well as western Canada.[16] Rising temperatures have already driven them to extinction in over one-third of their previously known habitat in Oregon and Nevada, while in the Great Basin (the arid region between the Rocky Mountains and California's Sierra Nevada) they have recently disappeared from eight of 25 mountain locations where they were documented in the early 1900s. Nationally, their situation is now so dire that the US Fish and Wildlife Service is considering the pika for protection under the Endangered Species Act. Some pikas living at lower elevation gain a survival advantage by eating their own faeces, which allows them greater access to nutrients in mosses that they can eat in cooler areas to avoid foraging in open, warmer areas for more nutrient-rich food.[17] But, despite their unusual strategy, it's possible that pikas will be the first mammals to become extinct in the US due to climate change.

The conifer forests of eastern North America are among the habitats being hardest hit by climate change, with some suffering up to 87 per cent mortality from changes stemming from the warming trend.[18] The mountain pine bark beetle is a well-known villain, having devastated conifer forests from New Mexico to British Columbia. With 88 million hectares of forest infested, and 70–90 per cent mortality rates for infested trees, these creatures, which are the size of a grain of rice, are rapidly altering entire ecosystems. The reason for this is warmer winters, which allow the beetles to extend their breeding season. And the problem is not confined to North America. Related species in the 200-strong pine-beetle family are devastating forests across the northern hemisphere.

Bushfires influenced by climate change are also permanently transforming some North American conifer forests into grassland, while drought and increases in water demand from heat-stressed trees are all adding to environmental mortality. As the forests die, so too will the myriad creatures that live in them.

There is one facet of the sixth extinction where climate change is not the sole culprit. Among the most dismal catastrophes to have struck the natural world in recent decades is the disappearance of many species of frogs and toads. About one third of all known 4740 species of frogs and toads are under threat, and in 2010 the International Union for the Conservation of Nature red list reported 486 species as critically endangered.[19] Up to 122 species are likely to have become extinct since 1980. Back in 2005 the cause of this calamity was unclear. Today, courtesy of new research, we know the spread of the chitrid fungus, which attacks the skins of amphibians, was the primary cause of many, but not all, amphibian declines. In *The Weather Makers*, I said that the extinction of Costa Rica's golden toad (*Bufo periglenes*) resulted from climate change. The latest studies support this, indicating 'medium confidence' (better than even chance) that climate change was the primary cause in this instance.[20]

So is the sixth extinction, as documented by Elizabeth Kolbert,[21] happening? In mid-2014 a detailed study was published that confirmed that the current global rate of extinction is about 1000 times greater than the 'normal' or background extinction rate.[22] The study makes the point that over the first decade of this century habitat loss was the single most important factor, while invasive species and climate change were also strong influences, but that as the century progresses, the influence of climate change will increase. It's still early days, but the climate scientists warning of overall species loss of 20 per cent or more as a result of the destabilising climate may yet prove to be correct.

CHAPTER FIVE

The Great Climatic Event Horizon

If looking into the sun may cause blindness, then human insights into nature entail a terrible price.

ANDREW GLICKSON, CONTEMPLATING THE GREAT CLIMATIC EVENT HORIZON, 2014

INCREASINGLY extreme floods, droughts, hurricanes, rainfall intensity and ice melts, and changing polar vortices: all are important and dangerous changes in Earth's climate system. But what of the system as a whole? Where is it heading? Will it settle into a new stable state and, if so, when? These are important questions, because only by answering them can we work out what action might be required in future.

The science of how the Earth system works and how it is adjusting to the build-up of greenhouse gases has gained clarity over the past decade.[1] Since the beginning of the Industrial Revolution at the beginning of the nineteenth century, the concentration of atmospheric CO_2 has been rising at an average rate of around 0.43 ppm (parts per million) per year. But that long-term average is a little misleading, because the rate of increase has sped up so greatly over the last 50 years. By 1998 the rate of increase had peaked at 2.9 ppm per year, but was still 2.89 ppm in 2012–13. The last 50 years have seen the fastest growth rate for atmospheric CO_2 in the entire geological record—faster even than that following the asteroid impact of 65 million years ago, which caused the extinction of the dinosaurs. The rate of temperature rise caused by the increase is also the greatest ever recorded in the geological record.[2] At current rates of increase, in 80 years global temperatures will have reached about 4°C above the pre-industrial average, creating climatic conditions last seen 55 million years ago, when an enormous eruption of methane from the sea floor caused a rise in global temperatures that lasted 100,000 years.

The impact of that ancient warming was truly astonishing. The tropics became largely uninhabitable, clothed in a thin, spiny vegetation unlike anything that had existed previously, while lemur-like creatures lived in the tropical rainforests that sprouted on Greenland. The best science suggests that the burst of greenhouse gas set Earth on a pathway towards being entirely ice-free. Were Earth to become ice-free again, sea levels would rise by 67 metres.[3] There is no reason to doubt that this would happen if temperatures were to increase by 4°C or more.

In a study that should act as a beacon for the world, scientists have recently tried to describe what Australia might be like if it were 4°C warmer.[4] They found that rising seas will necessitate a retreat from the present coast and the abandonment of much infrastructure, both in Australia's large cities and in smaller coastal communities, some of which may be obliterated. The heavily populated southern regions will be substantially drier and much hotter. Heatwaves, bushfires and storms will be more energetic—supercharged in a super-heated world. And of course there will be no Great Barrier Reef, no Kakadu wetlands or Ningaloo reef, nor any skiing or alpine daisyfields with their unique possums, frogs and insects.

We need do nothing to see this happen—it will simply occur as we go along with business as usual. And Australians alive today will see it: every year around 300,000 Australians are born, and increasing longevity means that many of those born this year will be alive in the twenty-second century. What will they think of us, who allowed such a world to be?

Scientists like Andrew Glickson, whose quote opened this chapter, think that, unless our trajectory changes, the twenty-first century will be marked by a 'climatic event horizon'. Just as the event horizon of a black hole marks a fundamental change in conditions, so will the climate event horizon mark a one-way trip into the unknown. The great question is to what extent we are already committed, by the powerful warming push of the greenhouse gases, to crossing it?

When I wrote *The Weather Makers*, I discussed three joker cards in the climate deck that might propel the world into abrupt, catastrophic and irreversible climate change: the collapse of the Gulf Stream, the destruction of the Amazon rainforests and a large-scale release of methane from the Arctic or ocean floor.

The Hollywood thriller *The Day After Tomorrow* popularised awareness that a collapse of the Gulf Stream could alter the climate. The Gulf Stream is a vast current of seawater, pushed and pulled on its journey north through the Atlantic by differences in density due to temperature and saltiness. In recent geological times, changes in climate have seen it cease to flow, and there are several ways that human climate change could affect it. By 2005 scientists at the Hadley Climate Centre in the UK had estimated the risk of the Gulf Stream collapsing this century as 5 per cent or less. Little additional science has been published since, suggesting that the risk remains small. The Hadley Centre has now archived its page on the Gulf Stream.

The Gulf Stream is a surface component of a larger system of ocean currents, known as the thermohaline circulation. Driven largely by differences in density, it transfers heat northwards from the equator. An important part of this system, known as the Atlantic Meridional Overturning Circulation (AMOC), has recently been shown to have suffered a 15–20 per cent decline over the twentieth century.[5] The slowdown was described as exceptional and without precedent over the last millennium. This is an important finding because the AMOC delivers about as much heat to Europe as the Gulf Stream does. The full implications of this slowdown are yet to be investigated, but it's clear that changes to Earth's heat transfer systems are being impacted by global warming, in some instances on a very large scale.

Destruction of the Amazon rainforests through drought-mediated climate change was the second potential catastrophe I explored. The rainforests of the Amazon basin are hugely important elements in Earth's climate system. They store unimaginably large amounts of carbon—if 80 per cent of it was released into the atmosphere, it would spike CO_2 concentrations from their current levels of around 400 ppm, to somewhere between 710 and 1000 ppm later this century. Without the forests, the surface temperature of the Amazon basin would be 10°C hotter than at present with 64 per cent less rain.

In 2005 the Amazon looked set to be hit with a double whammy. Climate change was threatening its rainfall, and the rate of clearing was fast enough to threaten the entire Amazonian ecosystem. The last decade has seen huge changes, including a massive reduction in forest destruction. So it's worth revisiting the issue of whether a collapse of the Amazon is still a real possibility.

It's a great curiosity that the Amazon forests make their own rain. Around 80 per cent of all the rain that falls there originates from water transpired through their leaves. The individual plants inevitably lose water when they open tiny pores in their leaves, called stomata, in order to take in CO_2. As the concentration of CO_2 increases, the plants don't need to open their stomata for as long, which reduces the water vapour released to the atmosphere and therefore the rainfall.

In January 2013 the UK Met Office published new projections on the Amazon. Factoring a combination of deforestation, fire and climate change into new, more sophisticated models, the Met came up with substantially different results from their earlier models. They found that as well as having substantial individual effects, deforestation, fire and climate change may combine in complex ways, and that events in one part of the forest could have effects that make other parts more vulnerable to the ongoing changes.[6] This new understanding of the complexity of climate and other human impacts on the Amazon forests has increased scientific uncertainty about how the Amazon basin might change. Two generations of Met Office global models gave very different projections: the first of extensive dieback from increased levels of CO_2 alone; the second, more sophisticated, model showed little forest change, except in the east of the basin, by the end of the century.

When the climate models disagree, it reminds us that there are limits to our ability to use our current science base to explain the highly complex climate system. Yet decision-makers must continue to frame responses and long-term policies to deal with the situation. In this context, summarising the risk to the Amazon rainforest from climate change, the Met Office says that its advice has not altered since the introduction of its new climate model. In other words, officials should continue to operate on the basis that the Amazon remains at risk.

One area of general agreement between the models is the prediction of large impacts in the eastern Amazon, and indeed major changes have been occurring there. Three major droughts (2005, 2010 and 2014) each of a severity expected approximately once in a century, have occurred in the region over the past decade. That these droughts are likely to have been influenced by climate change was reinforced by researchers based in the US, who concluded that deforestation was capable of decreasing rainfall and increasing drought severity.[7]

The Amazon forests normally absorb around two gigatonnes of CO_2 per year. In 2005 and 2010, as a result of the droughts, they turned from a carbon sink to a temporary supplier of atmospheric carbon, emitting an equivalent of around five gigatonnes of CO_2 during each drought event. And for years after the 2005 drought,

CO_2 drawdown was reduced. The drought that started in 2014 is not yet over, but already rainfall over São Paulo's catchments is the lowest ever recorded. For every month of 2014, rainfall lay outside the range of variation experienced since 1930.[8] In other parts of the Amazon, rainfall has also been well below normal. One hundred and forty cities in the vicinity have been forced to ration water, with conditions particularly severe in São Paulo State, home to one third of Brazil's population. In December 2014, São Paulo city's main water storage, the Cantariera Reservoir system, had only 7.1 per cent of its water capacity remaining.

Carlos Nobre, one of Brazil's most respected Earth scientists and climatologists, told the BBC in November 2014 that deforestation is to blame: 'There is a hot dry air mass sitting down here [in São Paulo] like an elephant and nothing can move it.'[9]

The third potential tipping point I identified in *The Weather Makers* concerned methane, particularly in the form of clathrates. Methane is currently responsible for around 20 per cent of the greenhouse gas influence, and emissions are growing.[10] Clathrates are an ice-methane combination. They abound in the depths of the sea, and in Arctic permafrost on land and underlying shallow areas of the sea. To stay stable, clathrates require cold and/or pressure. They exist in great volume in the world's ocean floor—there are between 10,000 and 42,000 trillion cubic metres of the stuff.[11] So vast is the resource that its partial release 245 million years ago is thought to have helped spike global temperatures by 6°C.[12]

In 2005 I argued that, of the three scenarios, the release of the clathrates was the least likely. This was in part because at that time the global atmospheric concentration of methane had been steady for several years and had even shown a slight decrease. This seemed to indicate that no large-scale release of methane was occurring. But by 2008 atmospheric methane concentrations had begun once again to increase, resuming a trend that had existed for most of the previous century. The cause of the methane 'pause' is not well understood. But clearly it was an anomaly.

There is no doubt that increased warming has the potential to cause huge methane releases. In its most recent report, the IPCC stated that potential emissions from melting permafrost are in the range of 50–250 gigatonnes over the twenty-first century under the highest emissions scenario, which we are currently tracking.[13] New understanding of the dynamics of the Arctic's methane reserves were gained during 2014 from studies both at sea and on land. Huge volumes of methane flowing from the sea floor in the Arctic were first detected in 2009.[14] But in February 2014 dramatic new insights and measurements were obtained when the icebreaker *Oden*, which had been specifically fitted for observing methane concentrations, began cruising off Siberia. Barely a week into its journey its crew of Swedish scientists reported vast methane plumes escaping from the seafloor of the Laptev continental slope.[15] It is still too soon to understand the full significance of these observations, which remain unpublished.

Meanwhile, on land, a serendipitous discovery by a helicopter pilot revealed another dimension to the methane issue. In July 2014, a mysterious crater, 30 metres wide, was spotted on Siberia's Yamal Peninsula. The air in the crater was found to contain exceptionally high concentrations (almost 10 per cent) of methane. Researchers suspect that the crater was caused by an explosive release of methane. In discussing the structure, Hans-Wolfgang Hubberten, a geochemist at the Alfred Wegener Institute in Potsdam, Germany, points out that 'over the past 20 years, permafrost at a depth of 20 metres has warmed by about 2°C, driven by rising air temperatures'.[16] If the methane that caused the crater was held in the form of clathrates, it probably originated from a depth of 100 metres or more, as clathrate deposits at shallower depths are rare. Even at such depths, the warming could well have destabilised it. Reindeer herders subsequently found other craters nearby.

Measurements taken in Siberia show that the amount of methane being released from the permafrost is five times greater than previously estimated.[17] Dr Jason Box of the Geological Survey of Denmark stated that if even a small portion of the methane in the Arctic escaped it could trigger irreversible, catastrophic climate change.[18] But we must remember that the reports thus far are anecdotal. Massive uncertainties exist in ascertaining methane emissions. Even the relative contributions of methane from human and natural sources to the global average figures remain uncertain, though it is a field of active research.[19] With large uncertainties remaining on the extent of current methane emissions from the Arctic,[20] it remains beyond the capacity of science to estimate the potential for a mass methane release from Arctic clathrates. Limited comfort, however, can be taken from the observation that global atmospheric methane concentrations continue to accumulate at around the same rate they have over the last few years.

Looking back over the past decade it's clear that the impacts of climate change have been every bit as severe and disruptive as scientists had predicted. But the impacts have not always been of the kind or scale envisaged. The large impact of the relatively small average global temperature increase on climate extremes is one example of this. From heatwaves to droughts and forest fires, we're already seeing changes that are destroying property and taking lives at a far larger scale, in many instances, than was thought likely just 10 years ago. With new, more powerful computer modelling, it's now clear that those impacts are set to grow in future.

Great advances have also been made in understanding how melting ice, particularly in Antarctica, might contribute to sea level rise. As a result many scientists are concerned that the worst-case scenario—that the world's oceans might rise a metre or more by 2100—is increasingly likely. If they prove correct, the cost in terms of money and disruption to human life from sea level rise is likely to dwarf all impacts felt from other changes thus far.

Ocean acidification, thought of a decade ago as a remote threat, is already having severe economic and environmental impacts. It may well turn out to be at least as disruptive as the warming trend itself. Climate change impacts are causing populations of vulnerable plant and animal species to decline as scientists predicted, though misinformation by climate-change deniers has confused public understanding in some instances, as have actions by governments that wish to ignore climate change. But new studies have quantified the rate of the sixth extinction, and explained in unprecedented detail its causes. Overall, the decline in biodiversity remains one of the most worrying trends of the twenty-first century.

On the positive side, some potential climate-change threats have either been downgraded or discounted. These include projections that the number of hurricanes experienced globally would increase and that the rainforests of western Amazonia might collapse. Furthermore, among the potential trigger points for a climatic event horizon, only one—a release of the clathrates—looms as an immediate potential threat. But too much comfort should not be taken from this—it only takes one trigger to change Earth's climate.

The progress in understanding how climate change is impacting Earth has been so rapid that I'm convinced that a decade hence we'll understand with far greater clarity what lies ahead. But a big factor in determining that future is the scale of greenhouse gas emissions. Will we reduce them? And, if so, how? That is the subject of the next part of this book.

2

THE KNIFE BLADE WE PERCH ON

CHAPTER SIX

The Great Disconnect

Ignorance is a dreadful thing and has caused no end of damage to the human race.

LUCIAN OF SAMOSATA, SECOND CENTURY CE

CLIMATE science advances by leaps and bounds, but in some countries politics lags far behind. Even in nations that lead in climate action, few politicians understand how dangerously and swiftly the burning of fossil fuels is altering our planet. Collectively, politicians are failing to act to maximise the chance of an acceptable outcome. A knowledge deficit among politicians is only partly responsible. Political lobbyists who set out to mislead are also to blame.

The ancient Roman lawyer Cicero urged us always to ask *Cui bono*? Who benefits? It is no accident that the climate sceptics hold greatest sway in the nations with the greatest investments in fossil fuels—including Canada, with its tar sands, and Australia, with its seemingly endless reserves of coal.

The response of sceptical politicians and officials when challenged on climate change is remarkably uniform. It usually goes something like this recent response from Tony Shepherd, Chairman of the National Commission of Audit of Australia, on national television on 24 June 2014:

> We're 1.5 per cent of the world's carbon dioxide emissions. I mean, if we halve it, it makes no difference. We shouldn't have been the world leader...Don't be the world leader when you're a carbon intensive country producing a miniscule amount of carbon compared to the rest of the world.[1]

The truth is that 180 individual nations out of the world's 193 are each responsible for less than 2 per cent of global emissions, which is precisely why it's important for all of those 'rounding errors' to work together. Furthermore, there is no evidence that leading on climate change damages the economy.

Those few nations that are leading are also thriving economically. Australia, for a brief period, was an example of this. When in 2012 it introduced its carbon tax, which taxed major polluters A$23 per tonne of carbon emitted, the country became a world leader on climate change action. This was a time of slow global economic growth, yet over the period the tax existed—1 July 2012 to 1 July 2014—Australia's GDP grew at a healthy 3 per cent per annum.[2] And the nation's carbon emissions dropped by 1.4 per cent.

Germany offers another example: it has experienced solid, long-term economic growth over the decades it has led on climate action. The German GDP growth rate averaged 0.29 per cent between 1990 (the Kyoto Protocol base year for emissions) and 2014, reaching an all-time high of 2.10 per cent for the second quarter of 2010.[3]

The idea, put about by the climate sceptics, that putting a price on carbon inevitably destroys economies is plainly wrong. Rather, we need policies, including a carbon price, that reduce dependence on fossil fuels. We all want those investments to be cost effective, but to deny the need for investment in combating climate change on the basis that it might 'damage the economy', as Canada's Stephen Harper did in mid-2014 when he decried Australia's 'job-killing carbon tax', is grossly misleading.[4]

Sadly, denial by politicians of the dangers facing society is nothing new. We've seen it all before—in the 1950s, when a technological revolution involving the power of the atom gathered pace at a fantastic rate, leaving politics far behind. Then, as now, the disconnect almost cost us our future.

Optimism soared at the dawn of the atomic age. The awesome power of the atom was intoxicating, and bizarre schemes to use it proliferated. Meteorologists proposed altering Earth's climate by exploding hydrogen bombs over the Arctic ice cap in order to melt it. American oil-men wanted to use nuclear weapons to mine Alberta's tar sands. A Hiroshima-sized bomb, detonated in the sands, they argued, would create a glassy flask by fusing the sand grains, at the same time as it altered the molecular structure of the tar, transforming it into sweet oil. In Australia, mining magnate Lang Hancock argued that A-bombs should be used to excavate harbours in the country's northwest so that iron ore could be more easily exported. Thankfully, either lack of capacity or public protest (in the case of the tar sands) prevented these follies.

Curiously, in Australia and Canada the pursuit of money and crackbrained ideas continue to dominate the political response to climate change. In those countries the gulf between what the scientists know, and what politicians choose to say, has only grown. At the same time, a sceptical media has helped create a gap in public understanding. Public ignorance has given the politicians the space to continue to fight last decade's battles, instead of addressing today's most urgent issue. The old and tired arguments of the sceptics, which have hardly changed in decades, ignore both scientific facts and lived experience.

Some companies bear a disproportionate share of the blame for this sorry state of affairs. The most important by far in the media is the Murdoch news empire, from Fox News to the *Australian* newspaper and the many tabloids that help shape public opinion. It's extraordinary to think that a media empire overseen by Rupert Murdoch, a man whose father was a frontline news reporter in World War I, continues in the twenty-first century to impede progress on this most vitally important issue.

How have the rearguard actions of the sceptics influenced things? If there ever was a defining moment in the political history of climate change it was the Copenhagen climate summit in December 2009. Efforts to destabilise the conference had begun months earlier, with secret plans to derail the meeting through a campaign to discredit climate science itself.

The conspiracy, which became known as the Climategate scandal, began with a sophisticated cyber-attack on the University of East Anglia's climate unit's server sometime prior to 17 November 2009. More than 1000 emails and 3000 other documents were stolen. The data was transferred to a server in Russia before selective, ambiguous and out-of-context lines from emails were posted anonymously from a Saudi IP address. When this was initially ignored, the conspirators contacted climate sceptics, urging them to whip up interest in the story. With their help a media frenzy was soon created, at the heart of which were allegations that the stolen material constituted evidence of massive scientific fraud.

Very little effort was made by the media to understand whether the stolen data was evidence of fraud, or not. Climate scientists seem to have been caught off-guard by the suspicions about their integrity that the leak caused, perhaps because, as historian Spencer R. Weart of the American Institute of Physics noted, 'We've never before seen a set of people accuse an entire community of scientists of deliberate deception and other professional malfeasance.'[5] And with the scientific basis of the talks under a cloud, the negotiations lost impetus.

An investigation by the East Anglia police failed to solve the email theft. As to any supposed wrongdoing by scientists, eight separate committees of investigation held in various countries and by a number of institutions found no evidence of fraud or misconduct. Tragically, it was all too late. The damage was done. Suspicion had been cast on climate science, and this weakened public confidence in the very basis of the Copenhagen negotiations, as well as affecting the resolve of those charged with negotiating a treaty.

Despite its unprecedented scale, environmental scientists with long memories would not have been completely surprised at Climategate. The publication of Rachel Carson's 1962 classic *Silent Spring* provoked a backlash from the chemical industry that was every bit as nasty. At the time Carson wrote, wholesale, largely unregulated use of pesticides was poisoning environments, species and even people across North America. The consequences, documented in *Silent Spring*, were horrific. The book resulted in legislation to ban or restrict pesticide uses. In an attempt to blunt the book's impact, chemical industry stooges shadowed Carson as she traversed the US, giving lectures disparaging her findings and discrediting her personally.[6]

Today's climate scientists are not the only researchers fighting bitter struggles against special interests. Some groups in the pharmaceutical industry have opened a war against scientists who oppose the veterinary use of antibiotics on the basis that their widespread use on animals will lead to the bugs that cause diseases among humans acquiring immunity. Embroiled in lawsuits, some researchers have lost their homes.

The conspirators did not triumph entirely at Copenhagen. Instead, decisive action by President Barack Obama yielded an unexpected payoff. Breaking into a meeting of the political leaders of China, India, South Africa and Brazil, the US president brokered a deal which became known as the Copenhagen Accord. Little more than a page long, it outlined an agreement in which nations set their own targets for emissions reductions by taking actions that suited their own particular economies.

Today, almost all that is being achieved politically at the global level stems from this one bold action. The steady accumulation of successful policy measures aimed at reducing greenhouse gas emissions that grew out of the Copenhagen Accord demonstrates that, from Vancouver to Beijing and on to Canberra, greenhouse gas emissions can be curbed at little cost. Indeed the majority of G20 countries have already decoupled their economic growth from growth in carbon pollution.[7] As a result, the political environment in the lead-up to the Paris climate summit in December

2015 is significantly different from that which prevailed prior to Copenhagen.

None of this means that the road forward will be smooth. As of early 2015, 'last gasp' deniers hold the reins of power in Ottawa, Canberra and a few other places, and they continue to turn their backs on sensible climate policy. Over the years they've gone from denying that climate change exists, to arguing that it isn't caused by humans, to saying that even if it is, it's too costly to fix. Even with this last crutch removed by experiences issuing from the Copenhagen Accord, the recalcitrants may yet form a 'coalition of the unwilling' in a final effort to frustrate action at the Paris summit.

Were this to occur, it could be disastrous for my country. Australia is highly dependent on exports in addition to coal and iron ore, among them premium wines, meat, dairy and other produce that trades on a 'clean, green' image for part of its value. Were Australia to gain a reputation as a recalcitrant polluter, markets might react. Consumers are more powerful than ever, and social media allows them to organise efficiently to express their displeasure.

It's worth looking more closely at Australia's current government, led by Prime Minister Tony Abbott, in this regard. Abbott took the leadership of his political party while the Copenhagen summit was underway—with a policy platform that included opposing measures to combat climate change. He immediately set about casting doubt on whether Australia's pledges to reduce emissions were of value by saying loudly and often that he would oppose many measures being developed by the government of the day. Nonetheless, he told the Australian public in the lead-up to the 2013 election that he would keep various measures already in place, such as a target for renewable energy. But, on gaining power, the new Abbott government set about dismantling almost all existing climate initiatives.

Its very first act was to abolish the Climate Commission, which I had been appointed to head in February 2011. The Commission's mandate was to provide the public with facts about climate-change science, economics and international action. The abolition was seen by many as an attempt to keep voters in the dark. Such was public concern, however, that five days after its demise the organisation was crowd-funded back into existence, and today runs as the Climate Council, an A$1.75 million a year not-for-profit institute, committed to keeping Australians informed about climate change.

When the Abbott government abolished Australia's carbon tax, coal burning for electricity generation spiked immediately. In the nine months immediately following the tax's abolition, emissions increases had erased all hard-won reductions in carbon emissions made over the last nine months of the tax's existence.[8] The Abbott government was also notable in being the first Australian government since 1938 not to have a minister for science. It abolished the Department of Climate Change and Energy Efficiency, and wanted to abolish the Climate Change Authority (which recommends targets and monitors the success of actions to reduce emissions), the Clean Energy Finance Corporation (which helps finance companies to reduce their emissions), and to render ineffective the Renewable Energy Target. But opposition from other right-of-centre political parties rescued those entities.

Abbott's unpopular war on climate-change action continues. His government tried desperately to keep climate change off the G20 agenda (which Australia hosted in November 2014), but the issue blew up when the US and China announced their pact to reduce emissions just prior to the meeting. Sadly, the Australian government continues to show little sign of constructive engagement in the lead-up to the Paris meeting.

In October 2014, the Global Green Economy Index, which rates countries on their improvement in quality of life and reductions in environmental impact, listed Australia in bottom place—behind Ethiopia and Rwanda—out of the 60 nations examined for performance on political leadership and climate change. The report attributed Australia's poor result for leadership to 'unconstructive behaviour in international forums'.[9]

CHAPTER SEVEN

Coal: Decline of a Giant

Whatever problem we are trying to fix, a lot of it is attributable to coal plants.

BRUCE NILLES, BEYOND COAL CAMPAIGN

THREE fossil fuels—coal, oil and gas—lie at the heart of the climate problem. Most coal is burned to generate electricity, and high-quality coal is composed almost entirely of carbon, making it the most carbon-intense fuel. Oil is composed of carbon and hydrogen, both of which yield energy, and is used primarily for transport. Gas is also composed of carbon and hydrogen, though it has less carbon relative to hydrogen than oil. It's a flexible fuel, used for electricity generation and transport as well as manufacturing fertilisers and chemicals. Fracking has brought huge volumes of cheap gas into the market, where it's competing with other energy sources, often with surprising consequences. And the new 'light, tight' crude oil produced by fracking has transformed markets. The following three chapters look at the state of the coal, oil and gas industries. All have seen astonishing changes over the past decade, and all face a challenging future.

The burning of coal to generate electricity remains the world's largest single source of carbon pollution. According to the International Energy Agency (IEA), things are going to get worse. In 2012 the IEA predicted that by 2017, coal use for electricity generation would rival oil as the world's largest energy source.[1] Given the energy mix then under development, enough polluting energy infrastructure would have been built by 2017 that, were it all to run for its full lifetime, humanity would be locked in to warming Earth by more than 2°C.[2]

The countries where the use of coal to generate electricity looks set to grow are generally so poor that their leaders feel they have no choice but to generate electricity at the lowest cost, and then suffer the health, environmental and climate-change consequences. India is probably the market where the coal industry's hope for a profitable future rides highest. One indicator is Coal India, the largest coal-producing company in the world, which is 90 per cent owned by the Indian government. After the election of the Modi government, Coal India's share price skyrocketed. But then, in Modi's first budget, India's coal tax was doubled (to US$3.40 per tonne), with the proceeds to go to installing renewable energy plants. The news for foreign coal suppliers has also been bad, with Piyush Goyal, India's new Minister of Power and Energy, planning to cease thermal coal imports by 2017.[3] With some policy at least backing renewable energy, coal's future in India remains as uncertain as it was prior to the Modi government's election.

Even some wealthy countries cling to coal. Turkey, for example, plans to build 75 new coal-fired power plants, despite the fact that on 13 May 2014 it suffered the worst coalmining disaster in the nation's history, with over 300 miners killed in an underground blast.[4] And Australia's dependency on coal looks set to continue, at least as long as the Abbott government remains in power. As a result of the carbon tax, coal's share of the Australian electricity market had shrunk to 69 per cent. But when the tax was abolished, coal burning increased to provide 76 per cent of the nation's electricity supply.

The immediate future of the global coal industry, however, will be decided in China, and there things are looking better for the environment. The true cost of burning coal, in terms of health and environmental impacts, is now clear for all to see, and a backlash against increasing coal use has been brewing in China for years. As a first step in cleaning up its air, China decommissioned its older coal-burning plants and developed policies for reducing its dependence on coal. Then, in 2013 it announced a cap on total coal use—a move the Australian coal industry claimed the Chinese would never be able to achieve.

A year later China had proved the Australians wrong: in the first half of 2014 its coal consumption actually dropped—albeit by a tiny fraction—for the first time this century.[5] Over the same period, China's GDP grew by 7.4 per cent. China now joins a long list of countries where coal consumption and economic growth are decoupled. China's coal imports also took a hit, rising by just 0.9 per cent, down from 15 per cent a few years earlier,[6] while domestic coal production dropped by 2.1 per cent, with another 2.5 per cent drop projected for 2015.[7] Chinese demand for coal has reduced to the extent that the China National Coal Association has called for a 10 per cent production cut. The government has recently increased import tariffs on foreign coal.[8]

These changes, swift and unexpected, have left many analysts baffled. The great question is whether this is a one-off, or the start of a trend. Analysts Wood Mackenzie believe that any talk of peak coal in China is illusory, and that demand will grow from four gigatonnes to seven gigatonnes per year by 2030.[9] Other commentators agree that coal use will continue to grow, citing increases in hydro-electricity as the reason that less coal was burned in 2014. But Chinese hydro-electricity generation has not departed from its long-term growth trend, indicating that the causes of the reduction in coal burning probably lie elsewhere. One important factor affecting energy use overall (and therefore coal use) is the growth of the Chinese service sector. It requires less energy per unit of GDP than manufacturing, so, as it makes up a greater proportion of overall economic growth, Chinese energy demand will slow. If reducing energy demand is indeed a factor, then the downturn in Chinese coal consumption may be here to stay.

Increasing energy efficiency is widely underappreciated in coal's decline, not just in China but worldwide. As energy prices have risen, measures such as LED lighting, increased use of insulation and improvements in appliance efficiency have become widespread. This has had a huge impact on demand in the developed world, which was long profligate with electricity because its price was so low. In Australia, for example, energy efficiency measures, along with household rooftop solar photovoltaics, have led to a decrease in the volume of electricity traded in Australia's electricity market every year for the last five years. It's a great illustration of seemingly small changes cumulatively having a big impact.

Australia is an interesting vantage point from which to watch coal's changing fortunes. The nation's coalminers and government agents were highly active as China built its coal-fired power plants. By exporting to China, Australia quickly became a coal export superpower, controlling a greater share of the global coal trade than Saudi Arabia did of oil. But over the past three years Australians have had ringside seats as this historic industry struggled for survival.

In 2009, when prices were rising and billions of dollars were being invested in new mines and infrastructure in Australia, an old coalminer told me that, after prostitution, coal is the second oldest industry in the country. It's certainly true that coal was one of the drivers of the colonisation of Australia. The First Fleeters of 1788 were instructed by the British government to keep an eye out for the black mineral, and they were not disappointed. By the time of Charles Darwin's visit in 1836 it was already evident that Australia would never run short of what Darwin called the 'motive force' of industry.

By the late twentieth century, coal was king. Lobbyists from the industry virtually dictated Australia's energy policy, and mine owners had deep influence in government.[10] In some states, corruption had also become entrenched. In 2013–2014 the New South Wales Independent Commission Against Corruption (ICAC) investigated the way certain coalmine leases had been acquired. The Minister for Primary Industry and Mineral Resources was found to have acted corruptly in awarding an exploration licence that benefited the family of Eddie Obeid, a former minister and a key figure in the New South Wales government. Obeid stood to gain

$50–$100 million as a result of the deal. Disturbingly, the ICAC found that 'with so many risks and opportunities for corruption, it was almost inevitable that corruption would occur at some point'.[11] Recently, the ICAC heard allegations that the developers of a new coal loader at Newcastle, north of Sydney, sought meetings with state ministers about the project after making financial donations to their party.[12]

Australia is not the only country where coal and corruption have gone hand in hand. In 2012, federal auditors in India found that the Indian economy had lost US$33 billion because coal-mining rights had been sold below their value. In 2014, almost all coalmining licences issued since 1993 were revoked by the Indian Supreme Court.[13]

The really big play in coal was supposed to be Queensland's Galilee Basin. The fate of the basin's coal resources is of global importance because the reserves are so large that their development is incompatible with the goal of keeping global temperature increases within 2°C. During the coal boom, nine separate mining proposals for the region were being developed. Five of them were for mines bigger than any then operating in Australia, and the projects overall were slated to produce more than 300 million tonnes of coal a year. That's around 1.5 times the total amount of thermal coal exported by all of Australia's existing mines, which would increase the world's total seaborne coal trade by almost a third.

The plan was shattered, beginning around 2011, by a falling coal price along with a reduced demand due to improved efficiencies and various air-quality and carbon-emissions regulations. Added to this was the plummeting cost of renewable technologies. Coal was in deep trouble.[14]

In the four years to March 2015, coal has lost half its value.[15] The price of Australian thermal coal has sunk to A$65.79 per tonne. Around half of the coal produced for export worldwide would be mined at a loss at that price. As of 2014, 10,000 coal workers in Australia have been laid off as coalmines are being idled or shut, or as mine productivity is ruthlessly pursued.[16] Some mining companies continue operations only because they have signed long-term 'take or pay' contracts—so the losses they make by mining and transporting coal are less than they would sustain if they shut the mine. Loss-making and even marginally profitable mines are grimly hanging on, hoping that someone else will go bust first, and the coal price will recover.

Between them, Anglo American, BHP Billiton, Rio Tinto and Glencore control 25 per cent of the global coal export business.[17] All these companies are highly active in Australia. BHP Billiton, Rio Tinto and Anglo American have now pulled out of the development of a crucial port for shipping coal from the Galilee Basin. Only two international investors remain—GVK and Adani—both Indian. In late 2014, then Queensland Premier Campbell Newman gave Adani a surprise gift when he announced that Queensland's taxpayers would help fund the 300 kilometre-long railway line required to get the coal from the Galilee Basin to the Abbot Point coal terminal adjacent to the Great Barrier Reef. The public was not to be told how much money would be contributed, nor the terms of the deal.[18] The public funding of up to half of the cost of the rail infrastructure clearly helped the Adanis unlock capital to fund the overall project.

But on 31 January 2015, in one of the most astonishing election results in Australian history, the Newman government was thrown out of office. The new Labor government stated that it would not invest in the Adani rail development. With circumstances changing rapidly, it remains uncertain whether the mines proposed by Adani and GVK for the Galilee Basin will be able to operate successfully.[19] Indian coalminers, incidentally, are not alone in facing massive capital risk. Worldwide, US$268 billion is slated to be spent on new coalmines by 2025.[20] But with the average cost of mining coal in the Galilee Basin projected to be about A$100 per tonne, the wide grassy valleys of Queensland's hinterland may remain a pastoral idyll—for the moment at least.

The contraction of Australia's coal industry has come at great personal cost to some. In 2012, the Australian Climate Commission held a public meeting in the heart of Queensland's coal country. In the audience of several hundred was a rugged-looking miner. He sat silently as we outlined the impacts that climate change was having on Australia. At question time his hand went up. He spoke quietly as he explained that he had once been a farmer, until drought and flood, both, as he now understood, influenced by climate change, had driven him from the land. He'd had no choice but to seek work in the mines. Was he, he asked, doing the right thing by his two daughters, aged eight and ten? None of us knew what to say, beyond that his first responsibility was to put bread on his family's table. And that his industry was likely to change in future. But governments surely have an obligation to ensure that new opportunities open for such people, through the creation of structural adjustment programs. I often think of that miner, and hope that new opportunities, perhaps in clean energy or carbon farming, are coming his way. But that hope depends in part on the deployment of new technologies and government encouragement of sustainable regional development. Sadly, Australia's governments, both state and federal, are lagging on both fronts.

Coal is in at least as much trouble in the US as it is in Australia. Given current trends, demand for coal in electricity generation in the US could decline by around a quarter (228 million tonnes) by 2020.[21] This is partly due to cheap gas, but also to better environmental regulation. Europe is also seeing a big decline in the use of fossil fuels for electricity production. In 2014, EU countries shuttered 13 gigawatts of coal and gas-fired electric plants, and opened eight gigawatts of new capacity. So the overall reduction was five gigawatts—about the equivalent of five nuclear power plants, or about a twelfth of Australia's current electricity demand. Over the next six years, Europe intends to shut 'most' of its coal and natural gas plants for electricity supply, seeking 'to set an example for developing nations from China to South Africa on how to reduce greenhouse gases blamed for climate change'.[22]

Despite all the setbacks faced by coalminers, a recent assessment indicates that the world's remaining viable coalmines continue to have the potential to trigger catastrophic climate change. As Carbon Tracker Initiative researchers put it: 'Plotting a linear decline in coal demand based on what is known to be feasible or part of policy today does not achieve a [world less than] 2°C [warmer than the pre-industrial average] world'.[23]

As the influence of coal declines, and alternatives become economically viable, the hitherto largely hidden costs of coal are being revealed. Air pollution is the most visible manifestation of its use, and lung cancer, heart failure, respiratory and kidney disease its legacy. Together, these ailments constitute the highest health burden in heavy coalmining areas.[24] Ascertaining just how many people die each year from the air pollution caused by coal burning is difficult. In the US, estimates of deaths caused by all electricity generation (of which coal burning is by far the most dangerous component) from 2005 to 2010 vary between 13,000 and 24,000 per year.[25] These numbers are lower than those from earlier decades, due in large part to coal-fired power plants being forced to install sulphur scrubbers. But the numbers remain appallingly high.

A 2008 Harvard Medical School study examined a wide range of costs resulting from coal use across the US economy. The annual public health burden in Appalachian communities (where much coal is mined) was estimated at US$74.6 billion. The cost of fatalities among the public due to coal transport by rail was $1.8 billion, and the health costs of air pollution from coal combustion were estimated at between $65.1 billion and $187.5 billion. Mercury impacts added between $414.8 million and $29.3 billion to the health bill. Subsidies to the coal industry cost between $3.2 billion and $5.4 billion, while the clean-up of abandoned mine lands added another $8.8 billion annually.

The cost of burning coal in terms of contributing to climate change (which are hard to estimate) was put at between US$20.6 billion and $205.8 billion per annum. The estimated total annual cost to the US economy of burning coal is thus between $175 billion and $523 billion. That equates to a cost of between 9c and 27c for every kilowatt-hour of electricity generated.[26] Even at their lower bounds, these costs are very large. It is astonishing that we have persisted with the fuel for so long.

CHAPTER EIGHT

What Future, Oil?

The use of solar energy has not been opened up
because the oil industry does not own the sun.

RALPH NADER

IN late 2013, I attended a dinner with tar-sands executives in Calgary, Canada. I had met earlier with the provincial government's committee on climate and energy, and I had given a talk on climate impacts in which I mentioned the possibility that the oil price might drop if demand decreased, as had recently been seen in the coal industry. The dinner was convivial, and towards its end I asked the executives what they saw as the greatest challenge to the fossil-fuel sector. For a moment there was a silence. Then a fellow who had said little all evening replied with a single word: 'Innovation'. Just how innovation is affecting oil is the central theme of this chapter.

Through the second half of 2014, the price of oil did indeed collapse, though not entirely as I had anticipated. By January 2015, sweet light crude was fetching just US$46.07 per barrel, its lowest price in nearly six years.[1] One factor was the rapid expansion of fracking in the US, which had increased the oil supply (fracking produces both gas and sweet light crude oil) by four million barrels since 2008, cutting US imports from OPEC countries by half.[2] Fracking is a new kind of mining whereby drilling extends vertically, then horizontally along gas-rich strata, typically shale. Following drilling, various methods, including the injection of chemicals, are used to fracture the shale or other rock in order to release the gas. The Saudi government, whose cost of production is far lower than that of the fracking industry, seem determined to kill this new source of competition by glutting the oil market.

If the price of oil continues to remain below US$50 per barrel, it will imperil parts of the shale gas industry (which uses fracking), in part because the oil derived from condensates in the gas provides most of the profit. The low price of oil is exacerbated by a characteristic of the shale gas industry, which distinguishes it from the traditional oil industry. Oil wells can produce for decades, but shale gas wells need to be re-drilled every few years. And even at $60 per barrel, the drilling of new wells in some shale oil plays is economically unviable.

Although in 2015 the event of 'peak oil' (that oil production will peak) seems distant, oil reserves that are inexpensive to exploit will eventually run short, and in the longer term the price of oil is likely to go up. The price may rise before that because many oil-producing countries will need to make up for lost profits caused by the low oil prices. Over the longer term they need more than US$100 per barrel to keep the nation's economy afloat. Moreover, nearly all of the big, unexploited oil deposits are in hard-to-reach, risky places like beneath deep water or the Arctic Sea. This makes them expensive to recover. Chevron's CEO, James Watson, has recently said that the break-even price for new oil development could be around US$100 per barrel.[3] So if the new oil deposits are to be exploited, a price above $100 must prevail. Any alternative, such as biofuels or electric vehicles, whose costs are lower, will price this new oil out of the market.

The oil industry did not anticipate the recent price drop; globally it is funding 163 exploration projects worth over a billion dollars apiece. In all, more than a trillion dollars' worth of investment, requiring a higher price than US$50 per barrel to yield a profit, is at stake. Over half of the projects require an oil price of more than $120 to become profitable.[4]

But there is another side to the oil price story, which relates to innovation and the demand for oil. One obvious influence on demand is the slowing of the Chinese economy. But I think that thousands of seemingly small initiatives, which are often overlooked, are also having an impact. From innovative financing that encourages building retrofits that reduce bills for heating fuel, to lighter building materials for transport vehicles that reduces fuel consumption, and on to the hybridisation and electrification of transport, investments are reducing demand for oil products.

Regulation, including America's strengthened Corporate Average Fuel Economy (CAFE) standards, is also driving down demand, as are new and innovative approaches to city planning. The rate and scale at which our cities are being transformed is, collectively, astounding. It's a two trillion dollar per year global industry. Almost everywhere the goals are the same: densification of housing, provision of new rail, light rail and bicycle-friendly infrastructure, and the reduction of carbon and other pollutants. All of this affects the relative economic cost and personal desirability of car ownership.

These changes all contribute to oil's declining share of the global energy mix, which is down from 46 per cent in 1973 to 31 per cent in 2012. They have been particularly influential in seeing the demand for oil peak, about 2005, in developed countries.[5] The extent to which declining demand has contributed to the recent collapse of the oil price is unclear. But some industry executives I spoke to thought that, at least in the early days of the price drop, oversupply and easing demand were contributing roughly in equal measure. To the extent that the demand for oil is easing as a result of efficiency measures and a switch to alternatives, then some of the same mice that are eating away at coal's future have entered the oil business. The mice may well be very small at present, but they are poised to grow rapidly.

Changes in the oil price have a far more profound impact on the industry today than they did in times past. Back in the 1970s, oil was cheap to extract, and there was little competition from renewable energy. Although high prices hurt the consumer, it mattered less to the industry whether oil sold at US$17 or $117 per barrel, because it cost less than $17 to produce, and there were no biofuels to compete with it when prices were high. There's an instructive analogy to be drawn between the oil industry's future and human anatomy. In the 1970s the fossil fuel industries resembled the arterial system of a fit young person. They could operate across a wide range of prices, just as a young person's cardiac system can operate across a wide range of heart rates. Today, however, the fossil fuel industries resemble a geriatric with hardened arteries. They can operate only within a narrowing band of conditions, and a spike, or an abrupt drop in price, is likely to see many of them off.

However, the opportunities for biofuel production remain limited. First-generation biofuels have been disappointing. Corn-based ethanol, for example, competes with food production, and because fossil fuels are used heavily at every stage of its production, it yields little, if any, carbon reduction benefits. When I wrote *The Weather Makers*, algae was showing great promise as a source of renewable fuel. Today, most innovations remain in the lab or operate at very small scale. Problems of contamination of pure algal strains with invasive species, the cost of centrifuging the algae to extract the product, and a 'boom and bust' cycle for cleantech in the US have all blocked algae fuels' pathways to scale. Not all hope is lost, however, for additional ways of making fuel with photosynthesis continue to be discovered. Some researchers, for example, are now looking at seaweed as a feedstock for biofuel.

The airline industry provides a case study of the problems faced by biofuels. The first non-petroleum based biofuels were used in commercial jet aircraft in 2008. Yet, as of 2014, demand from airlines remains low, so the manufacturers have been unable to raise the capital required to scale up their plants. The incumbent fossil fuels industry, which has already built plants that make jet fuel from crude oil, have a great advantage. [6]

Just why demand for jet biofuels remains low is complicated. There is little recognition from the public of the importance of biofuels, so little gain to be had for airlines wishing to brand themselves green. There are also many kinds of jet biofuels on the market, but which are best for the environment is not clear.[7] Until these constraints are overcome by a carbon price or some other mechanism, biofuels will remain less of a threat to the fossil fuel industry at times of high prices than they could be.

Outside the transport sector, the use of oil is being affected by the withdrawal of subsidies. In 2012, the G20 nations agreed to cut subsidies for fossil fuels, and their action is now having an effect. Fossil fuel producers were paid US$548 billion in 2013, a $26.5 billion decline on the previous year. Twenty-seven nations are now cutting subsidies for oil, coal and gas, and this will have a big impact, especially in the Middle East, where subsidised oil is burned to generate a third of the region's electricity, [8] and in Indonesia, India and Africa, where kerosene is still used for lighting and cooking. Alternatives such as solar lighting and efficient stoves using biomass or biogas are becoming cheaper and more widely available.

Another factor affecting the future of oil is the hitherto largely uncounted environmental and health costs that seem to result simply from the industry going about its business. Here's a summary of oil and gas disasters for just one month—March 2014—in a single country—the US.

> On March 13, eight people were killed in an explosion in East Harlem, New York; ancient, leaking natural gas pipes are believed to be the culprit. On March 14, an estimated 1500 gallons of oil were spilled from an Illinois pipeline and leaked into a ditch leading to the Kankakee River. On March 22, bunker fuel was spilled into Galveston Bay, a particularly important shorebird habitat. On March 25,

> BP spilled crude oil into Lake Michigan near Chicago. On March 31, a pipeline explosion at a supplemental storage facility for liquefied natural gas in Washington injured five people and forced the evacuation of 400 others.[9]

A certain level of minor spills and accidents have been hitherto seen by many as part of the cost of doing business—a bit like the air pollution effects of burning coal. But now that alternatives to oil exist, they are being seen as entirely avoidable catastrophes. Oil, incidentally, may be getting more dangerous to transport. Light, tight crude derived from the North Dakota Bakken shale is more volatile than conventional oil, and can detonate like a bomb. In July 2013, a train carrying Bakken crude derailed and exploded in Lac-Mégantic, Quebec, leaving 47 dead. Conventional rail, shipping and even pipeline infrastructure may not be appropriate for transporting such shale condensates.[10] If this proves to be so, an entire new generation of transport options may need to be developed, and paid for, adding to costs in an already competitive energy marketplace.

I have no illusion that the oil industry is likely to die anytime soon. But, like coal, its future is being challenged by a constellation of factors, which could see society using less and less of its products. That leaves just one fossil fuel industry with a future all but uncontested—gas.

CHAPTER NINE

Gas: Last Hurrah or Bridge to the Future?

When you've got nowhere to turn,
turn on the gas.

ANONYMOUS

ONE of the great geological controversies of centuries past was the battle between the Plutonists and Neptunists over the origins of Earth's surface. The Plutonists, who had Thomas Huxley on their side, asserted that rocks such as basalt and granite erupted in a molten state from deep within Earth, and that the other rock types, such as sandstone and slate, were derived from their breakdown and re-deposition as silt and mud. The Neptunists, who counted Goethe among their number, believed that Earth was originally covered in ocean, and that all rocks were formed as deposits on the floor of the ancient seas. By the mid-nineteenth century the matter had been all but settled in the Plutonists' favour. But then in 1912, Randolph Kirkpatrick, a curator of corals at the Natural History Museum in London, published a bombshell which re-ignited the debate.

In his book *The Nummulosphere*, Kirkpatrick argued that the entire planet consisted of fossilised fragments of extinct foraminifera of the genus *Nummulites*.[1] Foraminifera are amoeba-like organisms that live in the sea. Their supposedly universal distribution in rocks of all types was, Kirkpatrick argued, clear proof that the Neptunists were correct.

Kirkpatrick's idea is not as entirely as crackbrained as it sounds. He had noticed that the Egyptian pyramids were made of *Nummulites*' skeletons similar in size and shape to a dime (*nummulite* being derived from a Latin word meaning little coin). In fact *Nummulite* fossils abound in rocks across vast swathes of Asia, north Africa and Europe. But Kirkpatrick claimed that he could see them in basalts and granites as well—rocks in which no fossils had ever been found.

Many a scientist, after spending thousands of hours peering down a microscope at some repeated shape, is familiar with the phenomenon of seeing that shape *ad nauseam* on blank walls, and endlessly in dreams. Perhaps this is what happened to Kirkpatrick. Strangely enough, one of Kirkpatrick's colleagues, Otto Hahn—a German lawyer turned Swedenborgian and amateur petrologist—had been staring at other things. He claimed that all the world's rocks were in fact formed from the fossilised remnants of an ur-forest of algae. In accordance with his Swedenborgian beliefs, he asserted that algal fossils were present in meteorites, so the ancient algal forests must have originated in outer space.

Among the tiny filaments of the fossil algae, Hahn said he spied the remains of a minute triple-jawed worm. This he named *Titanus bismarcki*, in honour of the German Chancellor. Kirkpatrick was irritated both by the name and the challenge to his theory. But soon, the whole scientific storm in a teacup was forgotten, and the Plutonists again reigned supreme. I sometimes think of Kirkpatrick and Hahn when I read the works of economists, business leaders and politicians as they discuss the future of gas. We often see what we want to see, and nowhere is this more evident than among analysts and investors who believe that gas is our energy future.

Until the plunging oil price distracted analysts, the debate about gas played out endlessly in the financial pages of the world's newspapers. The extraordinary boom in the shale gas industry, beginning around 2000, seemed to open endless possibilities. Will gas replace coal? Will oil from shale gas extend the fossil fuel era for transport? And will cheap gas delay the adoption of renewables in energy generation? Shale gas already accounts for 40 per cent of US natural gas production and 29 per cent of oil.[2] As of 2014, most shale gas reserves were being drilled in the US. The sale of condensates alone provides a profit when the resource is exploited, so the gas can then be sold for next to nothing. This cheap gas has not only driven coal from the market, but helped rejuvenate the American economy, laying the basis for energy independence and the return to the US of large-scale chemical and manufacturing industries that had been moving offshore for decades. But the question of how big the future of gas might be remains.

In the early 2000s, at the start of the shale gas boom, gas was seen as a temporary energy bridge to a renewables-based future. Yet some analysts now believe that gas is here to stay. Oxford University's Dieter Helm, for one, sees a huge future for gas. In his 2012 book *The Carbon Crunch* he argues that gas will provide a bridge to a distant renewables future that is at least decades away.[3] His basic message is that gas will remain cheap and abundant in many parts of the world, while renewables such as wind and solar will remain expensive and constrained by their intermittent nature.

New energy technologies are developing so fast that it's already possible to test some of Helm's claims. First, is 'tight' gas (gas that typically requires fracking or a similar intervention to recover, of which shale gas is one type) widely available, and will it stay cheap? Wells drilled to exploit shale gas remain productive for only one to three years. The regular drilling of new wells adds to cost. Then there is the impact of the slump in oil prices, paradoxically created in part by fracking. As we've seen, it may threaten some fracking, and thus the expansion of production of this kind of gas. In addition, public resistance to fracking, particularly outside the US, has stalled the industry in several areas.

Then there are the unanticipated outcomes. In October 2014, the price of gas in Queensland (most of which is 'tight' gas from coal-seams) was close to zero.[4] That's because the industry was in the 'ramp-up phase'. It had production, but no capacity, as yet, to export its product. You would think that companies in this situation would simply turn off their production. But that has proven impossible to do because the wells flood and cannot be restarted. It's a surprise that will cost many millions of dollars, as otherwise useable gas is flared off or vented. So, while gas continues to boom, unanticipated problems are beginning to manifest themselves, dampening to some extent Helm's enthusiasm.

When it comes to renewables, the problems with Helm's predictions are even clearer. Helm seems to hate wind turbines as a blot on the landscape. And wind power, he says, is hopelessly uneconomical, with little prospect for any significant cost reductions or technological innovation. Yet in 2014 the world invested US$310 billion on installing clean energy, including a record $99.5 billion for wind, owing to several 'mega' onshore and offshore wind projects. In 2004 the world was spending just US$60 billion overall on clean energy. Investments have grown fivefold in a decade.[5] As we shall see when we examine the renewable energy industry, not only have innovation and cost reduction occurred on a massive scale in the wind industry, but they are accelerating sharply. According to Helm, solar is hopelessly expensive. But its cost has dropped, on average, to about twice that of coal, and in areas with good sunlight resources it is already cost-competitive with coal. Solar is anticipated by some to be globally competitive with coal by 2020.[6]

Those aware of the advances in clean technology have a different take from Helm, and are much less optimistic on the fracking craze. Dr Jiang Kejun, director of the Energy Research Institute/National Development and Reform Commission (ERI/NDRC), China, believes that the rush to gas is all about getting it out of the ground before it becomes uncompetitive with renewables.[7] It's not a view often heard in the US, but gas and renewables are already competing in the marketplace, and the most cost-effective technology will ultimately prevail.

So, is gas our bridge to the future? A spate of recent analyses predictably come up with different answers. Ezra Levant's *Groundswell: The Case for Fracking* is an unapologetic glorification of gas, arguing that fracking is the most important innovation of the twenty-first century.[8] Levant reveals that fracking has economic, geopolitical, environmental and patriotic aspects—which makes it complicated to evaluate objectively.

One attempt at a more objective approach is Bill Powers' *Cold, Hungry and in the Dark: Exploding the Natural Gas Myth Supply.*[9] Using a detailed analysis of gas-well performance, he argues that US gas is headed for a 'deliverability crisis' by 2015. In other words, there will be a gas shortage. But, whatever its immediate or even medium-term future, there's no doubt that gas is having a significant impact right now. A big enough impact, in fact, to cause some to doubt the short-term future of renewables. As David Crane, President of NRG Energy, which builds both gas and renewable power plants, said, 'cheap gas has definitely made it harder to compete', adding that only with renewable energy subsidies were companies able to propose wind projects below the price of gas.[10]

But continuous improvements in renewables are closing the gap, according to a panel of researchers at the Windpower 2014 conference:

> Advances in materials have allowed the design of longer turbine blades and rotors that can operate efficiently at lower wind speeds. Since 2012, a 'massive proliferation' of these turbines has driven average capacity factor increases up by 10 per cent at every level of wind resource. As a result of these advances, costs are falling; preliminary data shows that the average 2013 power purchase agreement was at $0.021 per kilowatt-hour.[11]

The economics of wind and gas are complex. Both have a global average cost of about US$84 a megawatt hour of generation capacity to install, excluding subsidies, according to Bloomberg New Energy

Finance. That's 3 per cent higher than the cost of a coal-fired power plant, and about half that of a nuclear reactor if government takes on the insurance and other risks. In the US, federal subsidies for wind are worth US$23 per megawatt. But gas projects can take advantage of complex taxation arrangements, known as 'master limited partnerships', that allow pipeline operators to pay less income tax. This effectively acts as a subsidy, which helps drive down the cost of gas.[12]

The economics of both wind and gas are affected by geography. The best wind resources in the US are in south Texas, where wind farms can be built for US$60 a megawatt hour, which is less than the $65 price of a high-efficiency gas turbine, according to New Energy Finance. But there are hundreds of other variables that help determine whether a utility invests in gas or wind.[13]

And there are other players. For example, in April 2014, Austin Energy, Texas, signed a 25-year purchase agreement contract with a 150-megawatt solar plant. The cost? Less than 5 cents per kilowatt hour. As one analyst reflected:

> The price reflects the benefit of the federal investment tax credit, but even without the credit the price would be 7 to 7.5 cents/kWh—still competitive with the utility's cost estimates for power from natural gas (7 cents), and well below the cost of coal (10 cents) and nuclear (13 cents) power.[14]

Without the tax credits, you might argue, Austin Energy would have signed up with a gas plant. After all, gas is half a cent cheaper. But there are good reasons for believing that solar represents a better deal, not least of which is the fact that solar runs on a zero-cost fuel stock. The gas market is notoriously volatile, and almost every analyst involved in the sector believes that costs will increase in the future.

Overall, in much of the US, in the short term at least, gas seems to have the economic advantage. Yet not all agree that the current situation is sustainable. Jeremy Leggett, green energy entrepreneur, notes that:

> We have learned that the top 15 players in US shale drilling have written off $35 billion since the boom started, and that investors are beginning to pull out. Meanwhile, production has peaked and is now falling in all but one of the major shale-gas drilling regions. The boom is looking like it could turn into a bust before too long.[15]

Outside the US, shale gas is an increasingly contentious issue, and that is slowing its development. Much of Western Europe, for example, looks set to reject the industry. In Australia some states have rejected it, while others have embraced it. China hopes for great things from it, but has made little progress exploiting it. Almost everywhere, farmers fear its impact on ground water. And, among younger people, there's a widespread perception that shale gas is 'fossil fuel-lite'—but still a fossil fuel—and so something they don't want.

Will the boom in gas make climate change better or worse? A recent study, using complex models, found that 'market-driven increases in global supplies of unconventional natural gas do not discernibly reduce the trajectory of greenhouse gas emissions or future climate change'.[16] In other words, gas is definitely not going to solve the climate problem. The models assume that the consumption of gas will have increased by up to 170 per cent by 2050. This might result in global greenhouse gas emissions in 2050 decreasing from what might otherwise have occurred, by 2 per cent at most. Or it might see them increase by up to 11 per cent. So, whatever else is claimed for it, we know that the gas boom will not alter our current trajectory from its worst-case scenario progress towards catastrophic warming.

CHAPTER TEN

Divestment and the Carbon Bubble

We see this as both a moral imperative and an economic opportunity.

STEVEN HEINTZ, PRESIDENT, ROCKEFELLER BROS FUND, ON DISINVESTING FROM FOSSIL FUELS, 30 SEPTEMBER 2014

A CLEAR understanding of the science of climate change has allowed the development of a carbon budget. Essentially, the calculation involves estimating how much atmospheric carbon will risk pushing average global surface temperatures 2°C or more above the pre-industrial average—a 'guardrail' that governments worldwide have agreed should not be exceeded.

The bottom line of the carbon budget is simple: to have a 75 per cent chance of avoiding more than 2°C of warming, over the first half of this century humanity can emit no more than 1000 gigatonnes of CO_2. That sounds like a lot, but by 2012 only 672 gigatonnes remained.[1] At the rate we're burning fossil fuels, we'll have used up the entire carbon budget by 2028—just over halfway into the budget period.

A big problem arises when we compare our remaining carbon budget with the world's valued reserves of fossil fuels (that is, those listed on stock exchanges worldwide). It turns out that if we burn all of the valued fossil fuel reserves, we'd release around 3000 gigatonnes of CO_2 into the atmosphere. As of 2015, our remaining budget is about 600 gigatonnes, which means that, if humanity is to have a fair chance of a decent future, about 80 per cent of the world's valued fossil fuel reserves must be left in the ground.[2] These excess fossil fuel reserves constitute 'the carbon bubble'. Some countries have larger 'bubbles' than others. Australia's known coal reserves, for example, represent about one-twelfth of the world's allowable carbon budget, a large part of which lies in the Galilee Basin.[3]

The recognition that fossil-fuel companies are fundamentally overvalued, because most of their assets cannot be used if we are to have a stable climate, has led to investors selling off their shares in various fossil fuel–based industries. Divestment started in the US in 2011, as a student movement on college campuses. It was given momentum by writer and activist Bill McKibben, the founder and leader of the protest movement 350.org, who took a speaking tour of the US in 2012 selling a simple message: 'If it is wrong to wreck the climate, then it is wrong to profit from that wreckage.'[4] In 2013, divestment gained significant global momentum with the publication, by the think tank Carbon Tracker, of a key report: *Unburnable Carbon: Are the World's Financial Markets Carrying a Carbon Bubble?*[5] It explains many of the key concepts behind the idea that stocks in fossil-fuel companies are overvalued because much of their asset base is unusable. Reviewing the material, oil majors Shell and BP agreed that the burning of the world's fossil fuels would lead to temperature increases pushing past the 2°C limit, and investors worldwide began to take the issue seriously.[6]

Six colleges, 17 cities and 12 religious institutions have already committed to selling their stock holdings in fossil fuels, and, as of October 2014, divestment campaigns continue at another 308 colleges and universities in 105 cities and states and at six religious institutions across the US. By September 2014, 181 institutions and local governments in the US and 656 individuals representing more than US$50 billion of funds had pledged to disinvest.[7] Their pledges were presented to UN Secretary General Ban Ki Moon as 120 heads of government (that's nearly two-thirds of all government heads) met in New York on 23 September 2014 to discuss climate change.[8] Among the disinvestors is the US$860-million Rockefeller Brothers Fund, which was built on the Standard Oil fortune. Their announcement resonated around the world.

The push to divestment has now gone global. In October 2014, the Australian National University announced its own disinvestment program, a move that sparked unprecedented criticism by members of the conservative Abbott government, including the treasurer, Joe Hockey. Investors, however, pushed back, saying that governments had no right to dictate how investors chose to deploy their funds. Much larger divestment plays will occur in years to come. In early 2014, the Norwegian government announced a review of investment strategies of its pension funds, and on 27 May 2015 Norway's parliament proposed rules to direct its largest pension fund to sell its assets in companies that earn at least

30 per cent of their revenue from coal. The rules gained bipartisan support and are expected to become law, with the sell-off beginning in 2016.[9] As of 30 June 2014, the total value of the fund, known as the Government Pension Fund Global, is US$889.1 billion. That's 1 per cent of global equity markets.[10]

During the first half of 2015, the global divestment movement grew spectacularly. The Church of England announced that it would not invest in companies dealing with fossil fuels, and 10,000 people signed a petition urging the Dutch pension fund ABP to disinvest. Asset managers claim there is a growing demand for investment products that have little or no fossil-fuel components. It is difficult to know just how much money is being invested in such strategies, but Gordon Morrison, a managing director at FTSE International, believes that about 80 per cent of institutional investors, including pension funds, are considering some sort of divestment.[11]

The latest research on carbon budgets attempted to establish just what types of fossil-fuel resources are unusable, and which countries they're located in.[12] It found that more than 80 per cent of known coal reserves, about 33 per cent of oil and 50 per cent of gas reserves must stay in the ground if we are to remain within budget. Analysing viability on a cost and location basis, the report found that the enormous coal reserves of China, Russia and the US must not be mined. Nor should the natural gas reserves of the Middle East, or any oil that may exist in the Arctic. Nearly three quarters of Canada's tar sands must also stay in the ground.[13]

The fossil-fuels industry asserts that investors simply selling shares in their companies does them no harm. But as a means of highlighting public disquiet about increasing carbon pollution, divestment is effective. It's arguably the most powerful challenge to the fossil-fuel industry's social licence to operate we've seen to date. The movement is spreading swiftly, and a corporation's 'carbon risk' is being taken seriously by more and more investors. As financial services company Standard & Poor's noted in May 2014:

> Investors are paying increasing attention to the impact of carbon and climate risk on corporate credit quality, yet their focus has largely been on regulated liabilities that reflect direct risks from regulations such as emissions trading schemes and other carbon pricing mechanisms. Outside of highly polluting industries, however, few companies recognise or account for the cost of carbon on their operations.[14]

Among the carbon risks that companies face are inflated values caused by including carbon resources as assets, when in fact the carbon resources can't be burned if we're to have a safe climate. They also include risks of increased carbon pricing, legal risks from health-related and other actions and reputational risk—for example, for banks that finance new coalmines and associated infrastructure.

In light of the new environment generated in part by disinvestment, Standard & Poor's recommend that each investor should 'examine the impact of carbon pricing on corporate credit from four risk aspects: environmental regulations, emissions market pricing, business risk across the value chain, and financial risk on profitability, cash flow, and asset and liability valuation'.[15]

There is a prospect that even fossil fuels once regarded as relatively green, such as gas, will be affected by carbon risk. As noted in June 2014 by Fatih Birol of the International Energy Agency, because the global energy-generation assets are getting greener as more wind and solar is built, unless carbon capture and storage is deployed to sequester CO_2 from gas-fired plants, by 2025 gas-fired plants will have higher than the average carbon intensity.[16]

Take addressing carbon risk a step further and you end up in green investment. Consulting firm McKinsey notes in a recent report that 'the quality and availability of sustainability data has improved' to the point where investors are able to go beyond simply not investing in polluting companies or industries, and that better returns can be had by investing in industries with the best sustainability practices.[17] They quote the research of three economists that suggest that sustainability initiatives can improve financial performance:

> The researchers examined two matched groups of 90 companies. The companies operated in the same sectors, were of similar size, and also had similar capital structures, operating performance, and growth opportunities. The only significant difference: one group had created governance structures related to sustainability and made substantive, long-term investments; the other group had not. According to the authors' calculations, an investment of $1 at the beginning of 1993 in a value-weighted portfolio of high-sustainability companies would have grown to $22.60 by the end of 2010, compared with $15.40 for the portfolio of low-sustainability companies. The high-sustainability companies also did better with respect to return on assets (34 per cent) and return on equity (16 per cent).[18]

Green bonds provide another avenue for investors. They are a way of raising finance to help solve environmental problems, including the climate problem. The funds are used solely for the stated purpose, and the loan repaid with either a fixed or variable rate of return. They are particularly attractive to institutional investors such as superannuation funds. The World Bank has issued more than US$3.5 billion in green bonds for climate-change-related matters since becoming the first organisation to provide the bonds in 2008.[19] The issuing of green bonds is expanding swiftly, and is likely to stand at $40 billion for 2014—a 20-fold jump from 2012.[20]

Whatever the fossil-fuel industry is saying about divestment, politicians in countries with abundant fossil-fuel resources are now taking the matter seriously. In March 2014, the UK government's Environmental Audit Committee voiced its concerns that the carbon bubble might be a threat to equity markets. The committee's chair, Joan Walley, told the BBC that:

> The UK Government and Bank of England must not be complacent about the risks of carbon exposure in the world economy. Financial stability could be threatened if shares in fossil fuel companies turn out to be overvalued because the bulk of their oil, coal and gas reserves cannot be burnt without further destabilising the climate.[21]

The concern is essentially that if fossil-fuel companies, being so very large, continue to be overvalued because of inclusion of unusable assets on their books, the economic shock of a mass devaluation may destabilise the economy. With the Paris climate meeting looming, the issue of the carbon bubble is becoming more urgent. As the UN's climate chief, Christiana Figueres, said early in 2014:

> Those corporations that continue to invest in new fossil fuel exploration, new fossil fuel exploitation, are really in flagrant breach of their fiduciary duty because the science is abundantly clear that this is something we can no longer do.[22]

If this is in fact the case, an orderly process of asset devaluation for unusable reserves of oil, coal and gas is clearly in the best interests of economic stability.

CHAPTER ELEVEN

Where's Nuclear?

All the waste in a year from a nuclear power plant can be stored under a desk.

RONALD REAGAN

NUCLEAR power plants can generate electricity without emitting carbon pollution. Calculated on the cost of electricity per kilowatt, nuclear power can also be cheap, especially if governments assume insurance risks and responsibility for waste disposal. A decade or so ago nuclear power was viewed by many, including myself, as having a role in the transition to a clean energy economy. It remains the preferred choice for many on the political right, who argue that the industry has huge potential for expansion, if only the greenies would get out of the way. But the future of nuclear power does not hinge solely on green politics. Economic factors will also play a large role.

According to the International Atomic Energy Agency, the number of nuclear power plants rose rapidly from the industry's beginnings in the mid-1950s, and global nuclear power-generating capacity peaked in 2010 at 375.3 gigawatts. By 2013 it had declined slightly to 371.8 gigawatts. If we look at the nuclear industry's share of the global energy mix, however, the story is starker: in 1996 nuclear power was providing 17.6 per cent of the world's electricity, but by 2013 only 10.8 per cent.[1]

It's worth comparing these numbers with the growth of renewables, particularly wind and solar. In 2000, renewables (including hydro power) contributed 18.7 per cent of global electricity generation capacity. By 2012, that number had risen to 22.7 per cent.[2] The generating capacities of wind and solar power are increasing at the same, rapid rate that nuclear power was in the 1970s and 1980s. By 2013, global wind capacity had grown to 320 gigawatts—equivalent to the capacity of nuclear in 1990.[3]

Nuclear power is not without its champions. China, with 26 nuclear power plants, has another 23 under construction. By 2020, these reactors will be producing 80 gigawatts of power. But that adds up to a mere 6 per cent of China's electricity supply. If the national building program continues as planned until 2030, China's nuclear power plants will then be generating 200 gigawatts of electricity.[4]

Pakistan also has grand plans for nuclear power, hoping to build 32 new reactors by 2050.[5] Given the state of Pakistan's economy, however, these plans are probably best regarded as aspirational. Russia's planned nine new nuclear reactors, to be built by 2017,[6] are perhaps more likely to eventuate. Most of the 31 nations with nuclear reactors have had mixed policies. A referendum held in Sweden in 1980 decided that the country should eventually close down all of its nuclear power plants, but in 2009 it was decided that 10 reactors should be retained. Canada was a nuclear pioneer, opening its first experimental reactor in 1947 and developing its own kind of reactor—the Candu (Canada Deuterium Uranium). Until around 2011, the country planned to expand its nuclear fleet. A new reactor was to be built in Alberta, in order to supply electricity to the tar sands, but market uncertainty has seen the plan dropped. Ontario, which hosts most of Canada's reactors, also planned to build more nuclear reactors, but declining electricity demand has led to a policy reversal, and Ontario now intends to shutter 2,000 megawatts of capacity.[7]

Among its traditional champions, nuclear power is falling out of favour. Its decline has been particularly steep in Japan, where problems for the industry began to multiply rapidly after 11 March 2011, when a tsunami led to the meltdown of all six reactors at Fukushima I Nuclear Power Plant, north of Tokyo. More than 300,000 people were evacuated, many of whom remain in temporary accommodation. A huge public backlash against nuclear power set in. The *Japan Times* reported: 'By shattering the government's long-pitched safety myth about nuclear power, the crisis dramatically raised public awareness about energy use and sparked strong anti-nuclear sentiment.'[8] By May 2012, Japan's last remaining nuclear reactor had been shut down, leaving the country entirely nuclear-free for the first time in 42 years. Then the Japanese government announced a ban on the construction of new nuclear power plants, and a 40-year lifetime limit on any existing ones that might be reopened.

The shuttering of all nuclear power meant the loss of 30 per cent of the nation's generation capacity.[9] A major negative short-term impact was that Japan had to import more fossil fuels[10] (valued at 10 trillion Yen by early 2015). At the time of the catastrophe, renewables were not as competitive with fossil fuels as they are today. The energy rationing following the shutdowns had some negative economic impacts. But they also had some interesting, arguably positive societal impacts. For example, office workers in insufficiently air-conditioned offices adopted less formal dress, shedding jackets and ties, while working from home became more acceptable.

The drive for renewables is now strong in Japan, and with costs coming down dramatically they are the natural choice for powering Japan's future. In 2012, I visited Panasonic's Tokyo headquarters and heard from its employees how those forced to leave their homes after the Fukushima meltdown begged for solar panels, resolute never again to use nuclear. Partially in response to such feelings, Japan committed to a major boost to its renewables program. In August 2011, the government passed legislation to subsidise electricity from wind and solar. Sales of solar panels rose by 30.7 per cent to provide 1296 megawatts by the end of that year. At the same time Japan commenced construction of its first offshore wind farm. In December 2012, the pro-nuclear Abe government was elected, but the public remains trenchantly opposed to nuclear power, and the growth in renewables continues.

The Fukushima disaster resonated round the world. In its wake, plans for nuclear power plants were abandoned in Malaysia, the Philippines, Kuwait and Bahrain, and were radically changed in Taiwan. But nowhere outside of Japan was the impact as great as in Germany. In early 2011, nuclear power generated 17 per cent of Germany's power.[11] But on 29 May, just over two months after the

Fukushima disaster, Angela Merkel's government announced that it would close all of Germany's nuclear power plants by 2022. As a first step, eight of the 17 operating reactors were permanently shut down. The global engineering company Siemens then announced its withdrawal from the nuclear industry, and a major rethink of Germany's energy policy was put in train. Germany's *Energiewende*, or 'energy turning', promises to make that country a world leader in placing low emissions technology at the heart of a highly sophisticated and industrialised power grid.

For all its dire consequences, Fukushima has not been the only problem dogging the nuclear industry. Costs, difficulties associated with managing an ageing fleet of reactors, and the need to diversify sources of power to increase energy security are also encouraging nations—including France, the undisputed champion of civilian nuclear power—to retreat from the technology.

In June 2014, energy minister Ségolène Royal announced that France would cut its dependence on nuclear power from a current 75 per cent to 50 per cent by 2025. Old nuclear power plants will be replaced by massive investments in solar and wind power, with 3000 megawatts of offshore wind farms (the equivalent of four nuclear power plants) to be built by 2020.[12]

Overlaid on this is the changing nature of electricity generation and distribution worldwide. Huge cost decreases in wind and solar, and the modular nature of these technologies, make it less difficult to raise the required capital. A few million dollars will buy you a wind turbine or a decent solar array, and these can be added to as demand or opportunity eventuates. Nuclear power, in contrast, is only cost effective at a massive scale—typically about two gigawatts.

Such a plant could cost at least US$15 billion to build in the US or Australia and take at least a decade to complete, and the financial investment would take 40 years to pay off.

Investors rightly ask what the price of electricity from wind and solar will be four decades from now. Add in the problems of insurance and waste disposal, and it becomes clear why nuclear power is only progressing in countries where governments will provide huge subsidies in terms of accepting insurance risk, taking on decommissioning and providing fixed-price power purchase agreements that run for decades.

On top of this, the issue of long-term storage of high-level nuclear waste remains problematic. Over the 60-year lifetime of an average nuclear plant, enough highly radioactive waste to fill an Olympic-sized swimming pool is generated. The waste will remain highly radioactive for several thousand years, yet not a single long-term, high-level nuclear waste facility exists anywhere on Earth.[13] Given these factors, and the recent shrinking role of nuclear power globally, it's hard to avoid the conclusion that the chances of a nuclear revival seem slender indeed.

CHAPTER TWELVE

Sunlight and Wind: Winning the Race

> Here is an almost incalculable power at our disposal, yet how trifling the use we make of it!...What a poor compliment do we pay to our indefatigable and energetic servant!
>
> HENRY THOREAU, 1848

A DECADE ago some economists were predicting financial ruin for nations that switched to clean energy. Economies would be destroyed by carbon taxes, trading schemes and investments in renewables, they said. One such was William Lash from the Center for the Study of American Business. He said that if America were to achieve the modest reductions in greenhouse gas emissions envisaged under the Kyoto Protocol, wage growth would drop by 5–10 per cent, domestic energy costs would increase by 86 per cent, family incomes would fall by US$2700 per year, and the consumption of fossil fuels would drop by 25 per cent (the equivalent of stopping all road, rail, sea and air traffic permanently).[1] But the US has since reduced its greenhouse gas emissions by about 11 per cent from their peak level in 2008, and the economy has grown.

Warnings continue to rain down from the denialists. Australians, for example, have been told by the fossil-fuel industry and its supporters for years that any price on carbon will cripple the economy. Yet after a carbon price of about $23 per tonne was put in place in the middle of 2012, the Australian economy grew healthily. Where they have not descended into the realms of fantasy, denialists have sulked their way into stubborn misrepresentation. In August 2014, Maurice Newman, who chairs the Australian Prime Minister's Business Advisory Council, alleged that California suffers poor economic performance as a result of investing in clean-energy technology such as wind and solar. He compared it unfavourably with Texas's economic performance, which he said was outstanding. Regardless of the accuracy of his economic analysis, he completely ignored the fact that Texas is a world leader in wind power, having more installed capacity than California, with 9.9 per cent of its electricity production coming from wind.[2]

In fact, renewables are now successfully competing with fossil fuels. For two years running more renewable energy, including wind and solar, has been installed globally than fossil fuel-based generation.[3] And now a major bank, HSBC, is warning that some fossil fuels risk being stranded by shifts in economics, innovation and concerns about climate change.[4] Wind and solar (photovoltaic, PV) power are profoundly disruptive technologies: ultimately, they cannot simply be bolted onto existing transmission networks and markets; instead they will transform them. Similar transformations have been seen in other industries. Kodak invented the digital technology that finally put it out of business, while newspapers are struggling to transform their revenue models as people move to real-time news from a variety of sources online. Yet this is only the beginning. A number of automotive manufacturers have announced that they will have commercial driverless electric vehicles available in the next few years. Petrol station owners and staff, mechanics and taxi drivers, among others, will feel the impact.

The disruptive power of renewable energy is most clearly seen in Europe, where wind and solar have reached their greatest penetration. In the last five years or so, European energy utilities have lost half a trillion dollars in asset value.[5] Part of the reason is that a new energy market is developing, one in which individuals, businesses and communities own their own power plants. You no longer need to invest a billion dollars to build a coal- or gas-fired power station, because renewables are scaleable.

Renewables do away with the concept of businesses and individuals as simple customers; instead, they transform them into 'prosumers' (producer-consumers) who compete with the electricity utilities at the same time that they buy from them. In response to this abruptly changed world, some utilities are trying to survive by selling the technologies, including solar panels and storage options, that ultimately compete with their core business as energy providers.

Renewables are also swiftly changing the way we distribute electricity. New, high-voltage, direct-current interconnectors are springing up to link wind resources with population centres. The economics of transmission are also altering: in Germany, which privatised much of its transmission network years ago, there are calls for some transmission assets to be returned to public ownership. There is no doubt that this will be difficult to achieve. Germany's *Energiewende* represents an attempt by government and industry to steer the full complexity of the energy revolution, but this is no easy task. The clean-energy genie is out of the bottle.

It is extraordinary to see how far wind and solar PV have come over the past decade. Towards the end of his presidency, Bill Clinton pledged that wind power would be contributing 5 per cent of the nation's electricity by 2020. By August 2014 wind was already generating 4.33 per cent of America's electricity. In 2008 the US Department of Energy announced that its aim was for wind to be providing 20 per cent of the nation's electricity by 2030.[6] With wind on the brink of a huge expansion, because it is now cost-effective against other forms of electricity generation in many regions of the country, 20 per cent by 2030 seems pessimistic.

Solar PV is also well ahead of projections. Australian scientists played a huge role in the development of silicon solar cells, yet as recently as 2007 there were only 7000 grid-connected solar households in Australia. Today, however, the country leads the world in household solar installation. In mid-2014 there were 1.3 million individual installations (as opposed to 500,000 for all of the US), meaning that 20 per cent of Australians now have the benefit of solar power.

An important reason for the massive growth in wind and solar is declining cost. The manufacturers of solar panels have cut production costs by more than 80 per cent in the past five years, and efficiencies throughout the production and installation supply chain are driving costs down even further. Installers, for example, no longer have to pay a visit to quote for a job. They can assess your roof using Google Earth, cutting $150–$200 in costs. Inverters, devices that change direct current to alternating current and represent one of the larger remaining costs in a PV system, are now a prime focus of cost-cutting. But financing, developing a customer base, and dealing with the paperwork associated with feed-in tariffs and approvals represent half of the cost of an installation in the US. McKinsey estimates that cost-cutting here has the potential to reduce the cost per watt of installation in the US from the current US$2.30, to $1.60 by 2020.[7] The overall impact of removing cost is seen in Deutsche Bank's projection that solar costs will continue to decline at the rate of 5–15 per cent per year between 2015 and 2017. If overall electricity prices increase by just 3 per cent per year, this will make solar cost competitive against fossil fuels in 80 per cent of solar's target markets.[8]

The wind sector is also cutting costs. The industry is currently undergoing a reorganisation from something resembling a cottage industry to a highly automated production line like that of the automotive sector. Technological advances—such as gearless wind turbines, the capacity to 3d-print repairs to rotor blades, and the containerisation of all parts of a windmill, including the tower, which allows for easy transport—are transforming the wind industry almost beyond recognition. This revolution is projected to cut the cost of electricity from wind by 50 per cent in the next five years.[9] With no fuel costs, and diminishing maintenance costs (gearless wind turbines have far fewer moving parts), wind is set to become a ferocious competitor in the energy sector.

The business of generating and distributing electricity has long been one of the least innovative practices in the industrialised world.

Energy utilities were (and occasionally still are) owned by governments and run by bureaucrats. Great coal-burning plants generated electricity, and a grid of wires distributed it. When wind and solar technologies came along, they were derided. They would never provide baseload (a reliable, constant supply equal to the minimum demand over a 24-hour period). 'The sun doesn't always shine and the wind isn't always blowing' was a commonly expressed view. I almost never hear it these days, perhaps because in some parts of the world a new 'wavy' baseload is emerging, powered by wind and solar.

One reason that wind and solar are disruptive is that their fuel is cost-free. Once the plant investment is made, and maintenance paid for, they continue to run with minimal cost. This means that electricity generated by wind and solar will always sell into the market first. This affects the continuity of demand for gas- and coal-fired electricity. For old coal plants this can be the end of economic viability. They're expensive to stop and start, some requiring huge amounts of diesel fuel to warm them up in a process that takes days. In some markets, if they continue to generate when their power is not wanted, their owners have to pay up to $10,000 a megawatt for disposing of their electricity into the grid. Gas plants are more nimble in switching on and off. But the fewer days they run because wind and solar are generating, the more marginal the business case for owning and maintaining them becomes.

As mentioned previously, solar power allows individuals to generate their own electricity and so removes demand from the market. This is proving to be disastrous for the companies that distribute electricity. They have fixed assets, by way of poles, wires and transformers, which must be maintained. The cost of this is shared among customers. As customers are lost, the cost per remaining customer rises, which drives more and more of them into buying solar panels.

The response of some utilities is to try to introduce a larger 'fixed' charge, a fee simply to remain connected to the grid. In response, Australian photovoltaic owners are beginning to organise politically, under the banner 'solar citizens'. The utilities seem bound to lose because, as storage technologies mature (battery storage is projected to halve in cost by 2020), price increases for a grid connection will simply drive more people off the grid and into solar with storage.

In a recent commentary in *Nature*, Jessika E. Trancik argued that some renewable energy technologies have crossed the Rubicon. Solar PV system costs have fallen 10 per cent per annum over the last 30 years. And an analysis of patents shows that the innovation boom that's driving the cost reduction is still gathering pace. Prior to 2000 the number of patents relating to fossil fuels and renewables were roughly equal in number. Since then, however, patents relating to renewables have quadrupled to over 4000 per year, far outstripping growth in those relating to fossil fuels.[10]

In places as diverse as Texas and Denmark, wind energy is now cost effective against coal in an open market. Trancik predicts that solar too will be cost-competitive against coal by 2020, at least in places where sunshine abounds. What might the future hold for clean-energy innovation? Trancik notes that 'any major energy transformation will involve stumbles'. Meanwhile, her students at MIT are 'pitching clean-energy ideas to start-up investors with a sophistication that was rare even among experts ten years ago'.[11]

It's clear that the market for renewables is far from saturated. Elon Musk's company SolarCity has recently acquired Silevo, a manufacturer of low-cost, high-output solar panels. SolarCity has committed to building a factory in New York State that will be capable of producing more than a gigawatt of solar power per year by 2016. This will be followed by construction of 'significantly larger' plants. Given that in 2014 solar panels were in oversupply, why the commitment? According to Musk:

> The sun radiates more energy to the Earth in a few hours than the entire human population consumes from all sources in a year. This means that solar panels, paired with batteries to enable power at night, can produce several orders of magnitude more electricity than is consumed by the entirety of human civilisation.[12]

If the solar industry is to generate 40 per cent of our electricity needs by 2050, more than 400 gigawatts of capacity would need to be installed each year for the next 25 years. With global production in 2012 at a little over 30 gigawatts, Musk and his team believe there's lots of room for expanded production. His reference to battery storage is also significant. And that leads us to the story of the electric car.

CHAPTER THIRTEEN

At Last, EVs

In 2003, no one was doing lithium-ion batteries for consumer cars.

J. B. STRAUBEL, TESLA CTO[1]

I AM embarrassed to say that EVs (electric vehicles) got short shrift in *The Weather Makers*, receiving less than a paragraph. By 2005 electric cars had been around for over a century—and had gone nowhere in the marketplace. Hydrogen fuel cells, or even compressed-air cars, were looking more promising. What a change a decade has made!

In 2014, 30 models of electric car were available. The current market leader is the Renault–Nissan Alliance, which had sold 176,000 units by August 2014. Tesla is in second place with about 50,000 units sold. Mitsubishi, BMW, BYD (a Chinese company), Ford and Volkswagen, to name a few, also have models in the market.

Tesla is at the forefront of EV innovation. In September 2014, it announced its biggest investment to date. Partnering with Panasonic, Tesla will build a 'Gigafactory' in Nevada that is capable of producing half a million electric-vehicle batteries per year. It will be the largest factory in North America. Construction is due to commence in 2016, and by 2020 it will be producing more EV batteries annually than were produced worldwide in 2013. As a result, vehicle battery costs are expected to drop by 30 per cent.[2]

The Tesla Gigafactory batteries could prove to be a game changer, not only for transport but also for stationary electricity supply. They would do this by feeding electricity into the grid from their batteries at times of high grid demand, and then charging up when demand is low. As energy expert Chris Nelder puts it:

> If Tesla's new 'Giga factory' can achieve its goal and slash the cost of lithium ion batteries…it would probably put the cost of owning an electric vehicle (EV) below that of a cheap, average gasoline-burner. Then, EVs could pick up real market share (they currently have less than one per cent of the US market). That would enable vehicle-to-grid (V2G) and vehicle-to-building (V2B) technology to become a real player in grid power, after years of languishing for lack of enough EVs on the road to make it effective. Just 100 electric cars parked at a 150,000-square-foot office building during the day could meet most of that building's peak demand, shaving off the most expensive hours of the building's power consumption. With widespread deployment, technologies like this could reduce the amount of expensive peak generation capacity that utilities need to build, and reduce electricity prices across the board.[3]

Battery technology has not been waiting for the Gigafactory. Already, enormous strides have been made in improving battery performance, as well as lowering costs. In 2014, J. B. Straubel, the chief technical officer at Tesla, said that 'battery energy density [a measure of energy stored relative to battery mass] has doubled over the last ten years and the curve is not starting to plateau'.[4] And the cost of batteries for EVs has fallen more rapidly than projected, from about US$1000 per kilowatt-hour in 2008, to $410 per kilowatt-hour by April 2014.[5]

In June 2014, Tesla made its patents for the charging of EV batteries public property, saying that it couldn't possibly manufacture sufficient charging stations to meet future demand. Currently, there are several options: from home-installed plugs that charge overnight to direct-current fast-charge stations that can recharge a battery in 30 minutes and are often located at highway stops to battery exchange facilities that can swap a battery in 15 minutes. The many billions of dollars invested in EVs by car companies across the spectrum attest to a widespread confidence that they have in the future of electric motoring.

In June 2014, France announced that it intended to enter the EV age in earnest—perhaps not surprising given that the market leader in the field is Renault–Nissan Alliance. The French government pledged to install seven million EV charging stations by 2030 and to make 50 per cent of its public-sector fleet purchases from EV models. It also announced a €10,000 subsidy for those trading-in a diesel vehicle for an electric one.[6] Despite this dramatic shift towards EVs, the market for electric cars remains nascent: in 2013 Tesla produced a mere 20,000 vehicles, and it was expecting to make 33,000 in 2014.[7]

The greatest agent of change will doubtless be China. The

nation's first EV manufacturer, BYD, produced just 2000 vehicles in 2013, but it looks set to make 20,000 (mostly plug-in hybrids which can run 70 kilometres on a fully charged battery) in 2014.[8] With reforms to green energy incentives, which cut the US$65,000 price tag for a BYD plug-in hybrid by a third, China plans to have half a million plug-in hybrid and electric vehicles (the majority of which will be imported) on the road by 2015 and five million by 2020.[9] It's the sheer scale of the Chinese market that makes it so important, and I have no doubt that the prices of electric vehicles will drop as Chinese manufacturing picks up, just as the prices of solar panels fell.

One hopeful sign for fully electric vehicles concerns the sales trends of plug-in hybrids as opposed to first-generation hybrid vehicles that power their electric motor from energy conserved in braking or by using the petrol motor. Plug-in hybrids are selling far faster at the same stage of delivery into the market as first generation hybrids, which gives hope that they can build a far larger market share, and faster.[10] This is important because a limited supply of batteries means that plug-in hybrids (which have smaller batteries) are likely to reach the Chinese market earlier than pure electric vehicles.

Some electricity utilities hope that electric vehicles will drive up the demand for power, and so give their languishing industry a second life. But such hopes are forlorn. Estimates from Germany indicate that, if every vehicle on German roads were electric, national electricity demand would rise by only around 20 per cent, a figure that would not vary greatly among developed countries.[11] And because electric vehicles might feed into the grid at times of peak demand, when electricity is most expensive, they may actually detract from the profitability of the utilities. In fact, electric vehicles are perfect for storing energy from renewable (intermittent) sources like wind and solar, and they are widely seen as enabling wind and solar to take a larger share of the electricity-generation market.

The low oil price prevailing in early 2015 has caused some to wonder whether sales of EVs might slow as a result. It is possible that some potential EV buyers might be persuaded that petrol or diesel vehicles represent good value. But it's also possible that potential new car buyers will think that the wild fluctuations in the oil price are likely to bring far higher prices in future as OPEC countries try to recoup their losses. As of mid-2015, the impact is unclear.

Despite the astounding pace of innovation seen in EVs, nobody expects them to dominate the road anytime soon. But, like solar and wind, electric vehicles are a disruptive technology. It's reasonable to assume that by 2020 electric vehicles will be eating away at oil demand without creating demand for coal or gas. They are the perfect storage system for renewables, and the synergy created with their widespread deployment may well provide the momentum for a decisive end to the fossil-fuel era.

3

FIGHT FOR THE FUTURE

CHAPTER FOURTEEN

Adapting?

Adapt or perish, now as ever, is nature's inexorable imperative.

H. G. WELLS

NO matter how swift our transition to clean energy sources, inertia in the climate system means that global warming will continue for some decades after CO_2 emissions peak. So even if emissions were to begin to decrease today, we would still face the challenge of adapting to climate change. Here I will highlight only some of the more enterprising examples of climate adaptation that spring from our seemingly endless ingenuity.

When it comes to adaptation it is important to understand that climate change is a process. We are therefore not talking about adapting to a new, known baseline, but to a constantly shifting set of conditions. This is why, in part at least, the US National Climate Assessment says that: 'There is no "one-size fits all" adaptation.' Nonetheless, there are some actions that offer much and carry little risk or cost. The National Climate Assessment report says: 'Climate change adaptation actions often fulfill other societal goals, such as sustainable development, disaster risk reduction, or improvements in quality of life, and can therefore be incorporated into existing decision-making processes.' And yet, the report notes, people often fail to take such actions because of limited funding, or policy impediments or legal restrictions.[1]

Around the world, at a local level, people are adapting in surprising ways, especially in some of the poorest countries. Floods have become more severe and damaging in Bangladesh in recent decades, with up to a third of the country submerged for some time every year. Mohammed Rezwan saw opportunity where others saw only disaster. His not-for-profit organisation, *Shidhulai Swanirvar Sangtha*, deploys 100 shallow-draft vessels that serve as floating libraries, schools, health clinics and gardens, and are equipped with solar panels, internet access and video conferencing facilities. Rezwan is creating floating connectivity to replace flooded roads and highways. But he is also working at a far more fundamental level: his staff show people how to make floating gardens and fish ponds to prevent starvation during the wet season.[2]

Elsewhere in Asia even more astonishing initiatives are occurring. Chewang Norphel lives in the Indian state of Jammu and Kashmir, where he is known as the Ice Man. The loss of glaciers due to global warming represents an enormous threat to agriculture—and therefore the survival of the people—in this mountainous region. This is because the spring melting of glaciers releases water just when it's most needed for agriculture. Without the glaciers, water will arrive in the rivers at times when it can damage crops. Norphel's inspiration came from seeing the waste of water that occurred over winter, when it was not needed. He diverted the wasted water into shallow basins where it froze, and was stored until the spring.[3] His fields of ice supply perfectly timed irrigation water. Having created nine such ice reserves, averaging 250 metres long by 100 metres wide, Norphel estimates that he has stored about 200,000 cubic metres of water. Climate change is an ongoing process, so Norphel's ice reserves will not endure forever. Warming will overtake them. But he is providing a few years during which the poor farmers of the region will, perhaps, be able to find other means of adapting.

Increasing Earth's albedo (its reflectiveness) can cool the planet. In the Almería area of southern Spain a proliferation of greenhouses (which reflect light back to space) has reversed the warming trend locally, and actually cooled the region. While Spain as a whole is heating up quickly, temperatures near the greenhouses have decreased by 0.3°C.[4] This example should act as a spur for all cities, which often suffer from heat island effects (as built infrastructure tends to retain heat). By painting infrastructure white, cities may more than offset the warming they currently experience.

On the other side of the planet from Spain, the glaciers of the Peruvian Andes have already lost about a quarter of their ice. The Pastoruri Glacier in northern Peru has retreated about 1.6 kilometres in the last 30 years, causing the closure of the region's international ski tournaments, which were the most important adventure-tourism business in the area. Loss of the glacier is also threatening the livelihoods of farmers.[5] Glaciologist Benjamin Morales believes that a temporary answer to the loss of ice could lie in sawdust and white paint. He remembered that ice used to be carried down the mountain in sawdust to prevent it melting, so he covered part of the receding tongue of the glacier with a six-inch deep layer. The rest of the glacier continued to melt, but a year later the sawdust-covered section remained frozen.

Encouraged by Peruvian inventor Eduardo Gold, local farmers around a mountain with a glacier, Chalon Sombrero, that has already fallen victim to climate change have begun painting the entire mountain summit white in the hope that the added reflectiveness will restore the life-giving ice. The paint is made from lime, egg white and water, and the painting process takes around a week per hectare. The entire area to be treated is 70 hectares. The outcome is still far from clear.[6] But the World Bank has included the project on its list of '100 ideas to save the planet'.[7]

More mundane forms of adaptation are happening everywhere. A farming friend I've known most of my life owns a property in western Victoria. Over five generations and 150 years the property has been too wet for cropping and has been used only to graze cattle and sheep. But during the past decade declining rainfall has allowed him to plant highly profitable crops. Farmers in many countries are also adapting like this—either by growing new produce, or by growing the same things differently. This is common sense. But some suggestions for adapting are not. When a lobbyist for the polluting industries argues that we've lost the battle to rein in carbon pollution and have no choice but to adapt, it's a nonsense designed to make the case for business as usual.

Humanity will continue to adapt to the changing climate in both mundane and astonishing ways. But the most sensible form of adaptation is surely to adapt our energy systems to emit less carbon pollution. After all, if we adapt in that way, we may avoid the need to change in so many others.

CHAPTER FIFTEEN

Geoengineering: a Way Out?

> There are all sorts of reasons why geo-engineering may prove impossible, either politically or scientifically...But to treat research into the subject as taboo on the basis that ignorance is a viable defence against folly would be a dangerous mistake.
>
> ECONOMIST, 11 DECEMBER 2014

UNTIL recently, the vast array of ideas for modifying global systems to address climate change—which range from directly interfering with sunlight, to extracting CO_2 from the atmosphere—have been discussed under the umbrella term 'geoengineering'. Increasingly, however, researchers are beginning to discriminate between proposals involving interfering with sunlight (solar manipulation) and most of the rest and restricting the term geoengineering to proposals for solar manipulation and a handful of other methods.

Among the earliest and most thoroughly documented of all geoengineering proposals is Paul Crutzen's concept of putting sulphur into the stratosphere to reflect sunlight. The idea involves the multiple roles sulphur plays in the atmosphere, and builds on work that led to Crutzen's Nobel Prize.[1] Revisited and refined over the decades, it's arguably now the most concrete and well-costed of all geoengineering proposals.

The basic idea is that if enough sulphur can be injected high into the stratosphere—about 15–18 kilometres overhead—it will stay there for a moderately long period, reflecting sunlight back to space, so allowing Earth's surface to cool. Crutzen thinks that the sulphur could be sent aloft using balloons or artillery guns. There are of course many considerations involved in such a proposal, but cost being an important one makes it a good place to start. Even before the potential wider costs of such activities are accounted for, Crutzen's proposal may not be cheap. In 2006, it was calculated to be in the order of US$25–50 billion per year (or $25–50 per affluent person on the planet).[2] More recently, much lower cost estimates for the core operation of injecting the sulphur of $2–8 billion have been obtained.[3] Costs which may be incurred as a result of changing rainfall patterns and other unintended consequences of reducing the amount of sunlight reaching Earth's surface are not included in this figure.

One recent innovation to shed light on the core costs and practicability of the Crutzen proposal is the SPICE project, run out of the United Kingdom. SPICE (Stratospheric Particle Injection for Climate Engineering) is essentially a test of a cost-effective delivery system for injecting reflective particles into the stratosphere. It was initiated in late 2011 and is due to deliver its first scaled-down prototype in 2015. SPICE deploys balloons on 25-kilometre-long tethers that also act as hoses to deliver the particles. Fresh water, rather than sulphur particles, will be used in the prototype. Obvious safety issues, including aviation hazards, are being evaluated as part of the project. If SPICE works and is cost effective, the possibility of Crutzen-style geoengineering will become very real.

Something like what Crutzen proposes occurs naturally when volcanoes erupt and inject vast amounts of dust and sulphur into the stratosphere, and such eruptions give an insight into the enormously complex ethics and governance questions that these literally top-down interventions raise. Among the most obvious and deleterious impacts are that rainfall can alter, crops can fail to thrive, and freezing conditions can affect human activities. But volcanic eruptions elucidate other potential, less-expected side effects of geoengineering as well. For example, we know that stratospheric sulphur destroys the ozone layer. And excess sulphur can impact on biological processes.

While volcanic eruptions are a useful partial guide to geoengineering outcomes, there are potentially important differences between geoengineering and volcanoes, which may result in different impacts. For example, the sulphur particles produced by humans would be much smaller than those from volcanic eruptions, and this and other differences add a level of uncertainty to how the sulphur will interact with the atmosphere.[4]

On the other side of the ledger, there is no doubt that stratospheric sulphur from volcanoes reduces Earth's temperature, so we can be reasonably sure that Crutzen-style geoengineering can achieve its intended effect. But it's important to recognise that such geoengineering does nothing in terms of the carbon problem, which will continue to grow unless the pollution stream is reduced. This means, for example, that ocean acidification will continue unabated regardless of how successful stratospheric sulphur injection is in reducing temperatures. And an abrupt termination of stratospheric sulphur injections may be more dangerous than the warming it stopped, because grave and swift changes in climate could follow as a result of the rapid warming.

Other means of cooling Earth by affecting sunlight have been proposed, such as launching reflective balloons or mirrors, or highly reflective nanoparticles into the stratosphere, or adding materials other than sulphur.[5] Soot particles, for example, could be released to create mild 'nuclear winter' conditions. In this case, Earth's albedo would actually decrease, because the atmosphere would be darker (as opposed to increasing with sulphur) but surface temperatures would, nevertheless, decrease. A relatively small amount of soot would be required, proponents argue. But soot is a carcinogen, and some may doubt the wisdom of dusting our planet with the stuff. All of these proposals share the same basic flaw—they fail to deal with the fundamental problem, which is too much CO_2 in the atmosphere.

The oceans have also been suggested as a venue for what many dub geoengineering projects. One of the most widely discussed approaches involves releasing iron or some other fertiliser, which is naturally in limited supply and so constraining the growth of plankton, into the ocean in order to promote plankton growth. Since 1993, 10 experimental releases of iron have occurred in various ocean areas, with mixed results. The idea behind the experiments is that plankton stimulated by fertilisation capture more CO_2 via photosynthesis, and that when they die they drift to the bottom, where the carbon they contain is buried and thus removed, for millions of years, from the carbon cycle.

The oceanographer who proposed the initial experiment, John Martin, famously said, 'Give me half a tanker of iron and I will give you another ice age.'[6] But experiments show that except in special circumstances most of the carbon captured by the algal bloom fails to sink to the sea floor and quickly returns to the atmosphere and ocean. Indeed, recent studies indicate that the amount of carbon captured is five to 20 times less than previously thought. Under optimum conditions, it might equate to one sixth of annual human emissions.[7]

It has recently been found, however, that if diatoms (plankton that grow a silica shell) are stimulated by fertilisation, there is a better chance that the captured carbon will sink.[8] Silica is a limiting growth factor for diatoms, so if both silica and iron are released in the right areas, there is the potential that under ideal circumstances vastly greater amounts of carbon might be taken out of the system—perhaps an amount equivalent to annual human emissions.

There are concerns that widespread iron fertilisation might have severe unintended consequences, including promoting toxic algal blooms, reducing plankton diversity and depleting oxygen in the ocean depth (if the sinking plankton rots there). As a result, the Convention on Biological Diversity (CBD) decided in 2008 that no further experiments on ocean fertilisation should be carried out in non-coastal waters until there was stronger justification and a regulatory regime in place.[9] As of early 2015, no global regulatory regime had been agreed. But neither this, nor the declaration of the Convention on Biological Diversity, has stopped the experiments.

The most recent iron fertilisation experiment occurred in July 2012, when a chartered fishing vessel dumped 100 tonnes of iron sulphate into the North Pacific off western Canada. The action was initiated by the Haida nation village of Old Masset, which has around 1000 inhabitants. The action was quickly described in the media as a 'rogue geoengineering scheme' orchestrated by US entrepreneur Russ George, who was accused of convincing the Haida to fund and carry out the project. This interpretation was proved to be untrue. The Haida had their own reasons for conducting the work, and they did not think of what they did as geoengineering. Indeed it appears that they were acting in a way similar to the people of Kashmir, who divert rivers to freeze and so recreate glaciers, or those in South America who paint mountains white to bring the ice back. All of these actions by small-scale communities are attempts at conserving a valuable local resource placed at risk by global change.

The Haida are salmon people. They have depended on the migrating fish since time immemorial, and the iron fertilisation was an attempt to revive the failing salmon stocks, which, due to overfishing, pollution and dams, have been in decline for decades. Salmon breed in rivers but grow in the sea, and the small creatures they eat depend on plankton. Because the Haida understood that iron fertilisation can make plankton bloom, they felt that iron fertilisation may help the salmon.

The project cost around CA$2.5 million, which the villagers lent to the Haida Salmon Restoration Corporation (HSRC). They hoped to recoup their money by selling carbon credits if the project was successful in sequestering carbon. At first the fertilisation looked to be highly successful. It promoted a spectacular plankton bloom that covered more than 10,000 square kilometres. But, sadly, there was no scientific involvement in the project, so nobody can say how much if any carbon was sequestered. Nor are there scientific studies of the experiment's consequences beyond what was incidentally collected as events unfolded.[10]

Because it flew in the face of the CBD ban on iron fertilisation experiments, the action by the Haida enraged some environmentalists, who challenged it in court. The HSRC faces up to 10 charges of environmental violation under Canadian law.[11] Proponents of the scheme are nonetheless encouraged. Two years after the event, which is enough time for the fish to grow, the 2014 northeast Canada salmon harvest was huge, increasing from 50 million fish the year before to 226 million. On the Fraser River, which had only once before had a take of more than 25 million, 72 million fish were caught.[12] The Haida believe that their experiment was a success. But in the absence of scientific data it's impossible to know whether the increase was a consequence of the fertilisation, or not.

The Haida iron fertilisation action has sharply divided public opinion on geoengineering. Some cheer it as an example of a successful human intervention to increase nature's bounty, while others warn of the inherent dangers of the approach and the lack of scientific evidence. Indeed geoengineering is a divisive topic. Possible ways of cooling the planet seem endless: from space sunshades, to white roofs and cloud whitening, they range from sensible no-regrets actions to those worthy of Dr Strangelove. In light of ocean acidification, geoengineering schemes involving the direct pumping of liquified CO_2 into the ocean depths, as has been mooted by some Japanese researchers,[13] must surely be put in the Strangelove category.

Despite their contentious nature and many downsides, it's worth asking how well such schemes might work. Early in 2014, a review published in *Nature* looked at the potential of many geoengineering proposals. Five methods were examined: afforestation, artificial ocean upwelling (for example, by using plastic tubes and wave power to bring cool water to the surface), putting iron filings into the ocean to encourage algae growth, putting lime into the ocean, and solar radiation reduction. The study concluded that:

> even when applied continuously and at scales as large as currently deemed possible, all methods are, individually, either relatively ineffective with limited (less than eight per cent) warming reductions, or they have potentially severe side effects and cannot be stopped without causing rapid climate change.[14]

In late 2014, three teams of researchers from Leeds, Bristol and Oxford Universities, working under the banner 'Integrated Assessment of Geoengineering Proposals' (IAGP), examined an overlapping series of options. Determining what the actual impacts of such interventions might be and accounting for downsides was 'really, really complicated'. The researchers admitted that they didn't like the idea of geoengineering, but were 'more convinced than ever that we have to research it'. The stakes are extremely high. All methods used to block the sun's rays are likely to affect rainfall, including potentially disrupting the Indian Monsoon. The researchers reported that 'between 1.2 and 4.1 billion people could be adversely affected by changes in rainfall patterns' potentially caused by solar blocking.[15]

If the technical aspects of geoengineering are daunting, they are dwarfed by the politics. Who will decide what Earth's optimum average temperature should be? And geoengineering may create winners and losers. How will conflicts be resolved? Clare Heyward of the University of Warwick runs Global Justice and Geoengineering, a project that promises to shed light on some of these issues. She studies 'how arguments for the development of geoengineering technologies might change when considerations of global distributive justice are foregrounded'. She is also examining whether geoengineering raises any distinctive challenges in terms of climate justice, and hopes to differentiate between the particular features and problems of individual geoengineering technologies.[16]

A much anticipated major review of geoengineering options was published in February 2015. It results from collaboration by the US National Academy of Sciences, National Academy of Engineering, the Institute of Health Sciences and the National Research Council.[17] The report authors point out up front that geoengineering is no substitute for emissions reductions. The review consists of two reports, which the authors advise should be considered separately. The first examines approaches and technologies, under the title 'albedo modification', that seek to reflect sunlight to allow the Earth to cool and have traditionally been called solar management or geoengineering.[18]

This report finds that albedo modification does not address the cause of human-caused climate change, but that it could be achieved unilaterally, would be relatively inexpensive to deploy and is likely to be effective almost immediately. However, the report also finds that it introduces novel global risks and that its abrupt termination may have severe negative impacts. In light of the dangers associated with albedo modification, the report recommends against large-scale attempts to deal with the climate-change problem using these methods. At current levels of scientific understanding, the report authors write, albedo modification could 'pose environmental and other risks that are not well understood and therefore should not be deployed at climate-altering scales'.[19]

The second report deals with technologies aimed at CO_2 removal, including carbon capture and storage, biochar production, ocean fertilisation and direct removal of CO_2 from the air using various techniques.[20] It notes that many of these approaches involve strengthening the mechanisms of the Earth system that maintains the carbon balance, and that many of these mechanisms have been weakened by human degradation of the environment. It states that one of the problems faced by all efforts to draw CO_2 out of the atmosphere is that if they succeed, large amounts of CO_2 will enter the atmosphere from the ocean as the systems move to equilibrium, slowing the efficacy of the intervention as far as global warming is concerned.

The report makes it clear that not all CO_2 removal measures are considered equal. Ocean fertilisation is singled out as being particularly risky because 'previous studies nearly all agree that deploying ocean iron fertilisation at climatically relevant levels poses risks that outweigh potential benefits.'[21] The report authors also point out that more research is required to determine whether any of these approaches can work at the scale required.

The combined American Academies reports' authors and organising institutions have the gravitas to allow their work to act as a foundation stone in the development of a future global agreement on protocols concerning large-scale projects, aside from mitigation, to deal with the climate problem. Without a protocol to regulate interventions, there is no prospect of humanity agreeing to deploy any of the options discussed, even if they are urgently needed. Single nations or groups may still act, as the Haida did in 2012, but any project that affects global conditions yet results from the decision of one or a few groups will remain highly contested and may even lead to conflict.

The most interesting part of the review is the volume documenting the basket of approaches and technologies on CO_2 removal. These approaches and technologies do not as yet have a commonly recognised, overall name. I suggest that, because they are neither emissions abatement nor geoengineering, but a distinctive third approach, they be called 'the third way'.[22]

CHAPTER SIXTEEN

The Gigatonne Challenge

The prediction I can make with the highest confidence is that the most amazing discoveries will be the ones we are not today wise enough to foresee.

CARL SAGAN

'THIRD-WAY' technologies recreate, enhance or restore the processes that created the balance of greenhouse gases which existed prior to human interference, with the aim of drawing carbon, at scale, out of Earth's atmosphere and/or oceans. It's what plants, and a fair few rocks, do.

The pathways life has evolved to draw carbon out of the atmosphere and oceans are extraordinarily complex, and involve the magic of quantum mechanics. Their ultimate power source is the sun, which photosynthesis uses to take carbon dioxide and convert it into energy and solid matter, including the necessities we call food and fuel. Over an immensity of time, photosynthesis has remade our world. A breathable atmosphere, stupendous stores of fossil carbon, in the form of coal, oil and gas, and a non-toxic ocean, are all legacies of photosynthesis.

Although plants are an important component of the third way, it is about far more than plants alone. Technologies now exist that allow us to draw CO_2 out of the air without the help of plants, and to make the captured gas into useful materials. Third-way approaches can involve the making of carbon-negative or carbon-neutral cement, changed management practices for livestock, and the manipulation of some natural geological processes, such as the weathering of certain kinds of rocks, as well as some very surprising new technologies.

My engagement with this fascinating basket of technologies and approaches began in 2007 when I received a phone call from Sir Richard Branson. He said that he had read *The Weather Makers* and that the book had changed his mind about the urgency of climate change. He invited me to Necker Island in the British Virgin Islands and confided that he was doubtful that humans could reduce emissions sufficiently fast to avoid catastrophe. He was planning to launch an innovation prize, calling for sustainable activities that have the potential to withdraw at least one gigatonne of carbon (about 3.7 gigatonnes of CO_2—those oxygen atoms are heavy!) from the atmosphere per annum.

A gigatonne sounds like an enormous amount, and it is. In round figures, one gigatonne of carbon is about one-tenth of the volume of carbon pollution humans currently emit annually. An increasing number of predictions of humanity's future emissions pathways suggest that Branson's misgivings about our ability to avoid a climate disaster were prescient. Indeed, it is widely anticipated that if we are to stop short of 2°C of warming by 2100, humanity will need to be drawing many gigatonnes of carbon out of the atmosphere annually well before the end of the century.[1] So, for the decades ahead, that's the scale society needs to be working towards.

The £25 million Virgin Earth Challenge (VEC)—at the time the richest climate prize ever offered—was born from those ruminations. Branson asked me if I'd be a judge, and I agreed, joining Al Gore, James E. Hansen, James Lovelock and Sir Crispin Tickell. Over the years I've watched in astonishment as the applications rolled in. They have fundamentally altered my perception about how we might respond to the climate crisis. Back in 2007 the technologies and methods presented were rudimentary. As we draw closer to 2020, and as humanity's carbon emissions levels continue to grow, some of the technologies encouraged by the Branson initiative have developed to look more and more like indispensable tools for our survival.

Moreover, it has recently become clear that drawing a gigatonne of carbon out of the atmosphere is just a beginning. The 2015 American Academies report on CO_2 removal states that 'reducing CO_2 concentration by 1 ppm per year would require removing and sequestering CO_2 at a rate of around 18 gigatonnes per year'. That's around 4.8 gigatonnes of carbon annually—almost five times the VEC target. Reducing CO_2 concentrations by 100 ppm (and so returning them close to what they were before the Industrial Revolution) would require removing around 1800 gigatonnes of CO_2—the same amount that was added by human activity between 1750 and 2000.[2]

More than 10,000 submissions to the Virgin Earth Challenge

have now been received. In 2011 these were pared down to a shortlist of 11 approaches. From restoring land and reversing ecosystem degradation to the capture of CO_2 using resins, all 11 fall into two fundamental categories—biological and chemical—according to the ways they extract carbon from the atmosphere and/or the oceans.

Biological methods involve the removal of CO_2 from the atmosphere or oceans via photosynthesis, and then storing the captured carbon in a variety of forms—from living forests to charcoal and plastics, or locking it deep in the earth's crust. Chemical removal options use the weathering of rocks, or artificial means, to capture atmospheric carbon, and then sequester the carbon in a variety of places, some of which overlap with those utilised by the biological pathways.

These two categories differ in a fundamental way. The energy required to drive the biological processes is essentially free, being provided by the sun, via plants. This is a great advantage, but its flip side includes fundamental limits: the rate and volume of carbon able to be captured through biological pathways is dictated by the biosphere and its wonderfully reliable but relatively inefficient photosynthetic process. Photosynthesis is only 1 per cent efficient (it uses only 1 per cent of the sun's energy available to it). Solar photovoltaic (PV), in contrast, can be 20 per cent efficient.

Quite apart from its limited efficiency, the biosphere is already bearing a heavy burden. We have cut down forests, polluted watersand driven to extinction many species. We are also placing great demands upon what remains of it for food, materials and space to live. And we continue to damage many of its unique and priceless ecosystems through ocean acidification and climate change, which limit our ability to use it for third-way approaches and technologies.

More ecologically grounded ways of managing and restoring the biosphere offer increasing promise as a means to reconcile the needs of the modern world while also restoring the natural world to function as it evolved to.[3] There is great promise for work in this field to increase the carbon carrying capacity of ecosystems and soils.[4] It can, however, be complex to work out what the right practices are in each unique set of species, soils and seas around the world, and to ensure effective stewardship of the system in question. Getting solid numbers behind the effects, such as the scale, longevity and permanence of carbon removal using biological approaches, is absolutely critical.

Any innovation can have negative as well as positive effects, depending on how it's deployed, and there are ways that the biological measures could do more harm than good. For example, if people went after aggressively carbon-sequestering monocultures such as corn or sugar cane, or if we relied too heavily on storing carbon in biological stocks that may be destined to degrade because of changes already locked into the climate system, the long-term consequences could be severe.

The chemical category of technologies differ from the biological ones in that they all demand energy from human energy systems, either via electricity or the direct burning of fossil and other fuels, at some stage in the process. This is expensive, and until low-carbon, renewable sources become widespread, it has the disadvantage of adding to the problem (by burning fossil fuels) that it is trying to solve. On the other hand, many of the chemical technologies offer the advantage of both storing the carbon securely and/or creating something useful to humans in the process.

The option almost everybody thinks of when it comes to removing carbon from the atmosphere is to grow more trees. Trees, like all plants, grow by drawing in CO_2. They are, in fact, little more than congealed atmospheric CO_2, with half of their dry weight being composed of carbon drawn from the air. Many of the world's forests have been cleared or degraded in recent centuries, so restoring them offers considerable potential for sequestration. But the scale of reafforestation required to draw down a gigatonne of carbon is staggering.

Trees grow over a long time, and start out very small, so we must take a 50-year time horizon as we think about this option. Over half a century, we would require an estimated 3–7.5 million square kilometres of land to be reforested (7.5 million square kilometres is roughly an area the size of Australia or the contiguous states of the US). We would need to complete between 70,000 and 150,000 square kilometres (approximately half the area of the United Kingdom) of plantings each year, if we were to sequester on average a gigatonne of carbon annually for fifty years this way.[5] And the trees must be sustained for a century or more if the carbon is to stay out of the atmosphere for a useful period. Compared with many other third-way approaches, planting and maintaining trees is relatively cheap—reafforestation at this scale is estimated to be a modest US$20–$100 per tonne of CO_2 captured and stored. But at $100 per tonne, it would still cost $370 billion to sequester a gigatonne of carbon by planting trees. Costs may well come down with scale, but there can be no doubt that they will remain substantial.

Of course, planting forests can bring other benefits, from water catchment protection to reducing erosion and the conservation of biodiversity. But there are potential downsides as well. As pointed out recently in *Nature*, reafforestation at this scale can have unintended consequences by changing Earth's albedo.[6] Trees absorb more heat energy than do paler grasses, and this increase in heat energy captured at Earth's surface may offset, or even more than offset, any gains in reducing temperatures made by drawing all of that extra carbon out of the atmosphere. That's not an argument against trees and their multitude of benefits, of course, but if we are reforesting for climate-change abatement, such factors need to be considered.

Carbon is stored directly in soils, in the form of humus, charcoal or the root mass of grasses. The amount and type of soil carbon can be influenced by land use. For example, cell grazing (where livestock are kept in a dense flock or herd, and moved from one paddock to another after grazing a small area for a day or so, mimicking the grazing patterns of the ancient roaming herds) may result in increases in soil carbon, partly by favouring perennial species with large root masses. The practice can also increase the stocking rate, potentially allowing soil carbon to be stored and a greater profit (and thus at a negative cost). Other methods of storing carbon in soils, however, can cost as much as $100 per tonne, at which price it would cost $370 billion to sequester a gigatonne of carbon. Prices, of course, are expected to decrease as industries scale up. But, nonetheless, cost is, at present at least, a formidable barrier.

Currently, there is a lack of precise science on how much carbon can be stored in soil using livestock management techniques and for how long. For the moment, investing in such programs for carbon storage is an uncertain business—a bit like buying a bag of rice without knowing how much rice the bag actually contains or its use-by date.

It is also possible to burn waste biomass, including garbage and forestry and farm waste, then capture the CO_2 created, and store it deep in rock strata, a process known as Biological Carbon Capture and Storage (Bio-CCS). The only operational, commercial-scale Bio-CCS project in existence is the Midwest Geological Sequestration Consortium (MGSC). Located at the University of Illinois in the US, it is co-funded by government and industry. MGSC captures CO_2 from ethanol production, which results in a relatively pure stream of CO_2 compared with other types of biomass energy production (such as biochar). This minimises the cost of separating out the CO_2. The captured and purified gas is then compressed and injected into a deep saline aquifer, where it is hoped it will remain permanently. The program is only economically feasible because of government research-and-development grants. A permit from the EPA is required, and because the process is ground-breaking technology, such permits are not easy to obtain.

The amount of carbon generated by burning biomass depends on many factors. But about half of dry plant matter is carbon, and 3.7 gigatonnes of CO_2 includes 1 gigatonne of carbon. So you'd need to burn at least 2 gigatonnes of bone-dry feedstock such as sawdust to generate 1 gigatonne of carbon storable as CO_2. Costs very much depend upon the future development of CCS technologies, but are likely to be higher than US$100 per tonne of CO_2 stored ($370 per tonne of carbon stored). A cheaper alternative may be to store biomass as structural elements in buildings, or to bury it where it won't break down and release CO_2.

It's worth looking at technologies based on processing trees and other land-based plants—to pick them apart a little in order to understand what might really be achieved—because humanity has more experience of this approach than any other.

Before the oil industry developed to its mammoth size, many of the products it generates today were extracted from wood. Indeed, for more than a century prior to World War II wood chemists had been creating myriad useful chemical products from timber. Lye, saltpetre and potash are age-old products derived from wood ash. They were used in cleaning, food preserving and explosives manufacture. Charcoal was also a useful fuel product, and the early wood chemists found that the process of making it yielded various valuable complex chemicals. The wood was burned in a low-oxygen environment (a process known as pyrolysis) and the fumes were captured and condensed in a copper still, and then decanted through a series of barrels and pipes. The result was a range of fuels, solvents, explosives, dyes, antifreezes, preservatives and early plastics such as bakelite. By the 1930s, wood chemistry had developed to such an extent that the industry was producing versions of most of the products that we derive from fossil fuels today.[7]

The impacts on the forests were substantial. An average wood chemistry plant in the northeast of the US consumed 11,232 cords of wood per year (a cord is 128 cubic feet, or 3.62 cubic metres)—enough timber, if stacked 1.2 metres high, to occupy 3.34 hectares.[8] Forests were cut on a 30-year rotation. Shortly after World War II, the industry collapsed due to competition from fossil fuels, which were becoming available in bulk, and were transformable on a massive scale into fuels, fertilisers and plastics.

The climate problem has sparked renewed interest in wood chemistry. In 2008, a new research institute was set up within Hamburg's Johann Heinrich von Thünen Institute, and various companies around the world are experimenting with this old technology. The production of alcohol is emerging as significant. The word alcohol has an intriguing etymology. It comes from Arabic—*Al Kohl* meaning 'the powder', as you might guess, from 'kohl', which has been used as eye makeup for thousands of years. But alcohol does not exist in powdered form. The link comes because alcohol must be distilled and, when distilling essences, the old alchemists often found a powdery residue in the bottom of their flasks, which they named *Al Kohl*, and which was later transferred to the distillate.

Today, a form of alcohol—methanol—is much used as a transport fuel. First-generation biofuel technologies source it from food products such as corn, and the carbon benefit over fossil fuels is marginal at best. But methanol made with second-generation technologies is now under development. Utilising the wizardry of the wood chemists, methanol will be derived from cellulose, resulting in a far larger carbon benefit (because tree farming takes less fossil fuels than growing corn).

The production of charcoal (biochar) remains the principal aim of many involved with modern wood chemistry, and from a climate perspective, biochar is the most important wood-chemistry product. Biochar is a relatively pure, mineralised form of carbon, so it rots very slowly in comparison with wood. If mixed in with the soil, or stored in old mines, it can be a secure carbon store for a century or more. As in traditional charcoal-making, the biochar process begins with heating vegetable matter (typically wood or wood offcuts) in the absence of oxygen. This separates the carbon-rich charcoal from the other compounds, which escape with the steam. It takes little energy, and is considered to be a carbon-negative technology because it allows for long-term storage of carbon that was captured from the atmosphere by plants. In recognition of its importance in addressing climate change, the technologies that produce biochar, and work through its complex interactions with the soil equivalent of a coral reef, are among the leading 11 technologies and approaches chosen as finalists in the Virgin Earth Challenge.

Biochar can be added to soils in ways that give additional benefits, such as moisture and nutrient retention. But the product takes many forms, depending on what it's made from, and the temperature and speed at which it's made. Some forms of biochar are good for some soil types, but other forms, particularly those made at high temperatures, can be toxic. Biochar science is complex—getting consistent quality of biochar out of a kiln and matching its characteristics with particular soil types are serious challenges.

The first-ever report summarising the state of the biochar industry was published in early 2014. Its chief finding is that the industry is in a very early stage, consisting mostly of small businesses in Europe and North America, selling biochar products locally for gardening and tree care. On average only 827 tonnes are sold per year, and because the price of biochar has yet to benefit from the economies of scale delivered by mass production, it remains relatively expensive—about US$2.50 per kilogram. Biochar companies range in scale from cottage industries to small-scale industrial concerns. The most important feedstock is waste from forestry, which tends to be centralised (for example, around sawmills), and therefore easy and cheap to collect.

Barriers to the industry's expansion include a lack of financing, early stage technologies and a lack of demand for the product. The report states that education of stakeholders—from farmers to regulators and lenders—is the industry's most urgent challenge.[9]

There are, however, other quite fundamental barriers. Given variation in soil types the biochar may be stored in, feedstocks and processes, it is not yet possible to predict how much carbon will be sequestered and for how long. This limits the ability of farmers using biochar to enter carbon markets. These constraints may be overcome soon, however, because scientific research into biochar continues to expand rapidly. The number of peer-reviewed biochar-related publications increased nearly fivefold between 2009 and 2014, with over 380 papers published in 2013 alone.[10]

Biochar would need to be made and stored on a truly massive scale to meet the gigatonne-scale ambitions of the Virgin Earth Challenge. For example, theoretically, all of the agricultural and forestry waste in the world would need to be made into charcoal, as well as the yield of 100 million hectares of land growing energy crops, to sequester just one gigatonne of carbon per year. The challenge of mass production is compounded by the fact that agricultural waste is something of a misnomer. Much of the material that could be made into biochar is already used on farms to feed animals, as fuel for stoves, or as a soil conditioner.

Production costs presently range from zero to about US$60 per tonne, depending on feedstock and context. There are so many feedstocks, and so many soil types, that it hardly seems possible that biochar production could ever become a highly industrialised process producing gigatonnes of uniform and, therefore, inexpensive product. On the positive side, the biochar market is segmenting in interesting ways. Full Circle Biochar is a company that has focused on the consistent production of certain kinds of biochar, in the hope that the technologies it develops can then be licensed to large-scale manufacturers. The Biochar Company, in contrast, has focused on brand creation. Its product, Soil Reef, is tailored for the sequestration of carbon and the agricultural benefits of biochar.

Other companies produce biochar as part of wider operations. Phoenix Energy, for example, designs and builds small-scale (0.5 to 2 megawatt) power plants fuelled by biomass. They turn wood waste, agricultural waste and other biological waste products into a burnable gas, which is used to generate electricity and produce biochar, which is sold to farmers and gardeners, providing an additional revenue stream.

Cool Planet Energy Systems focuses on biofuel production, with the biochar by-product again providing additional income. With US$100 million in equity, it is one of the more mature companies in the biochar space. The biofuel that Cool Planet Energy Systems produces can be blended with petrol. The greenhouse gas saving comes from the fact that waste that would otherwise end up as CO_2 is converted into biochar and a fuel that competes with fossil fuels. The company claims that its product performs as well as petrol, and can be sold at a similar price. In common with most other kinds of biochar, their biochar by-product can be stored in agricultural lands, where it can help retain biodiversity, and enhance soil moisture retention. In 2014, broadcaster CNBC listed Cool Planet Energy Systems, based in Denver, Colorado, as one of its top-50 disruptor companies.[11]

For all the advances being made by the various companies, biochar remains far from a planet-saving technology. The problem is scale. In order to sequester a gigatonne of carbon, long-term, in the soil in the form of biochar, up to eight gigatonnes of dry biomass may be needed. This is in part because biochar slowly degrades over time. The rate at which it degrades depends on soil and moisture conditions, but even in the best case only a portion of the carbon fixed as biochar during the pyrolysis process will be sequestered for more than a century. The industry has a long way to go before it is contributing significantly to the production of biofuels and the sequestration of carbon. But, as one of the most mature industries in a nascent field, it's important in our fight to stabilise the climate. Indeed in a decade or two it may have transformed into one of the leading solutions.

So far, all of these options are land-based. When we turn to the waters, both the potential and the uncertainty increases. An important fact to recognise, however, is that the marine technologies can offer an almost immediate fix, at least in limited areas, to the ominous problem of ocean acidification.

Wetlands are spectacularly good at capturing CO_2 from the atmosphere. Wetland plants grow fast, and the oxygen-poor conditions in many wetland sediments are ideal for storing carbon.

Indeed many fossil fuels were formed in wetlands. Unfortunately, the world's wetlands, from mangroves to saltmarshes, have suffered enormous degradation in recent centuries. Many have been converted to dry land for a variety of uses, including pasture and cropland, golf courses, canal estates and industrial areas. In addition, some are lost inadvertently due to developments upstream, such as dams, which starve them of their water supply. But some degraded wetlands could be restored. Just how much CO_2 could be captured, and for how long, by restoring wetlands remains highly uncertain. Estimated costs of wetland restoration vary hugely—from US$10 to $100 per tonne of carbon captured.

The most exciting, if least understood, of all options concerns the cultivation of seaweed. While still very much a frontier prospect, if indeed not just beyond the horizon, the potential of this approach is enormous. The stupendous potential scale of seaweed farming was outlined in 2012 by Dr Antoine De Ramon N'Yeurt of the University of the South Pacific and his colleagues. They explain that because seaweed grows very fast, seaweed farms could be used to absorb CO_2 very efficiently, and at a large scale. The seaweed could be harvested and processed to generate methane for electricity production or to replace natural gas, and the nutrients left could be recycled.[12]

Their analysis shows that growing seaweed could produce 12 gigatonnes per year of methane, while storing 19 gigatonnes of CO_2 that result from the methane production. A further 34 gigatonnes per year of CO_2 could be captured if the methane is burned to generate electricity.

> These rates are based on macro-algae forests covering 9 per cent of the world's ocean surface, which could produce sufficient biomethane to replace all of today's needs in fossil fuel energy, while removing 53 billion tonnes of CO_2 per year from the atmosphere...This amount of biomass could also increase sustainable fish production to potentially provide 200 kilograms per year, per person, for 10 billion people. Additional benefits are reduction in ocean acidification and increased ocean primary productivity and biodiversity.[13]

Many of the technologies required to achieve this are already in widespread use, if at a comparatively minuscule scale. Indeed, seaweed farming covers hundreds of square kilometres off the coast of China alone. The required methane digesters are also a simple technology, which is widespread in agricultural use to transform waste such as piggery effluent, and could easily be utilised on floating factories.

Where would all the CO_2 be stored? N'Yeurt and his colleagues give four possible options, one of which is Carbon Capture and Storage in the ocean floor. Importantly, as we shall soon see, CO_2 storage in marine sediments at depths below 3000 metres is looking like a much more favourable prospect than geosequestration on land.

In its scale, efficiency and potential advantages, seaweed farming may provide spectacular solutions to the climate crisis. Covering 9 per cent of the world's oceans with seaweed farms, and then processing the voluminous product yielded, is far beyond our current capabilities. It is heartening to think, however, that no new technologies are required for this approach and that, by itself, it comes close to being able to negate our current global emissions.

CHAPTER SEVENTEEN

Silicate Rocks, Cement and Smart Chemistry

Cement manufacturing is a major source of greenhouse gases. But cutting emissions means mastering one of the most complex materials known.

IVAN AMATO, *NATURE*, 2013

THE chemical means of capturing and storing CO_2 are many and spectacularly varied. Direct air capture involves exposing 'sorbents' to the air, so that they absorb the CO_2. The gas can then be compressed and stored in deep rock strata. Swiss company Climeworks has developed a mobile CO_2 capture device that can remove one tonne per year of CO_2 directly from the air around it using a new kind of recyclable chemical filter system. When the filter reaches capacity, the CO_2 is driven off by heating the filter to 95°C. The CO_2 stream released is very pure, making it a commodity with some value. The filter can be re-used many times, and the fact that about 90 per cent of the energy needed can be supplied by low-temperature heat makes the technology potentially relatively low cost.

The Canadian company Carbon Engineering has a broadly similar approach; it intends to use captured CO_2 to make low-carbon fuels. Their experimental unit operated for about six months in 2012, and captured two tonnes of CO_2. Carbon Engineering is yet to construct a complete pilot plant, but anticipates doing so by 2017.

Scale is a formidable problem for these types of technologies. About 10 million shipping-container sized units, using 4–5 exajoules (a quintillion joules) per year of electricity, would be required to sequester one gigatonne of CO_2. By comparison, in 2008 humanity used 474 exajoules of energy, so this investment represents about one 150th of total human energy use.[1] No commercial demonstration plant yet exists, and cost estimates are hugely varied, from US$20–$1000 per tonne of CO_2 captured.

Another option is chemical capture of CO_2 from the air by exposing silicate rocks to weathering. The proposal involves accelerating the weathering process that occurs in nature by breaking large rocks into smaller pieces to increase their surface area and thus rate of weathering, and exposing them in conditions where they will weather quickly. Olivine is a beautiful, glassy dark green mineral from deep in the Earth's crust, which is transformed by naturally occurring chemical reactions into a variety of common rock types, including serpentinite. Grind olivine or serpentinite into a sand, and lay it on a beach, and it can continue to absorb CO_2 for years. But between 3.6 and 5.1 gigatonnes of rock is required to sequester one gigatonne of atmospheric carbon. That's a lot of rock to mine, crush and transport. The estimated cost is US$24–$123 per tonne of CO_2 removed.

Derbigum, a roofing company with factories in Europe and the

US, offers an interesting twist on this approach. It has developed a roofing product with a layer of olivine that reacts with rainwater to remove and permanently store atmospheric CO_2. Dutch company Greensand offers yet another solution: an olivine-based, carbon-negative lime-replacement product (for soil remediation) for use in home gardens.

Many seemingly far-fetched applications have been suggested for olivine. They include constructing olivine hills as monuments or public recreation areas, using olivine to make artificial reefs for tourism and fishing purposes, as an additive in fire-fighting, to both improve fire-suppression and capture carbon, through to using olivine sand as a replacement for conventional sand in beach replenishment. Olivine grains might also be used to enhance the growth of diatoms in biofuel production or for constructing building ventilation systems that will also control CO_2 levels during the day. Olivine-rich soils might be used to grow plants which hyper-accumulate nickel, or for producing magnesium carbonate spring waters similar to those that naturally occur in springs across Europe near olivine deposits. It's even been proposed that olivine-based carbon-capture devices be installed on ships. Located in the exhaust of the ship's engines, they would capture the CO_2 emitted and turn it into a carbonate that, if released into the ocean, could lead to the sequestration of additional amounts of CO_2 from seawater.

Olivine is not the only rock that can be used to capture CO_2. Lime produced from carbonate rocks such as limestone can also do the job. Lime production, however, requires considerable heat, and between five and six gigatonnes of rock would be needed to sequester a gigatonne of atmospheric carbon, at an estimated cost of

US$79–$159 per tonne. Because of the high costs involved, and the fact that no useful material is created in the process, the widespread use of olivine and other rocks for sequestering CO_2 will require incentives, as well as much research and development aimed at lowering costs.[2]

Carbon negative cements represent another chemical pathway with huge potential. The production of Portland cement, which dominates the cement industry today, generates one tonne of CO_2 for every tonne of cement produced. As a result, cement manufacturing contributes about 5 per cent of our current greenhouse gas emissions. Researchers are trying to understand the maddeningly complex chemistry of cement-making in order to find a way to reduce greenhouse gas emissions. One option being investigated is the use of a lower roasting temperature during production, which would involve burning less fuel. Another is to incorporate fly ash, a by-product from burning coal for electricity, in the cement-making process, which one company claims makes its cement carbon neutral.[3]

But it turns out that there are ways of making a cement that actually absorb and sequester carbon over long periods. These include a cement-curing process being developed by Solidia Technologies in the US that takes CO_2 from industrial waste and incorporates it into the cement. Solidia claims that its cement can be used to produce concrete that is stronger, more durable and more flexible, and that it costs less than conventional cement.[4] It uses the same raw materials and equipment, but less water, energy and time. Their product is now in the commercialisation stage.

About four billion tonnes of cement are produced worldwide

each year[5] at a market value of US$300 billion, and cement and concrete products are worth about $1.3 trillion annually. There is certainly the potential for such products to make a difference to the atmospheric carbon budget at a gigatonne or near-gigatonne scale. But to sequester a gigatonne of carbon using this technology, 80 per cent of the world's cement production would need to incorporate carbon-negative processes. One huge impediment to the uptake of these new kinds of cement is the risk-averse nature of the industry. Nobody wants their building or bridge to fail, and until the novel products have stood the test of time, and a global price on carbon emissions is introduced, it's hard to see them being adopted at scale.

Carbon-negative plastics offer a potential solution for storing CO_2 captured from the air. One of the leaders in the field is Newlight Technologies in California. As a result of 10 years of research, Newlight has invented and commercialised a carbon-capture technology that combines air with methane-based greenhouse gas emissions to produce a material they have called AirCarbon. It is a carbon-negative material that can be used in place of oil to produce a range of plastics that can compete with oil-based plastics on both performance and price.[6] From chairs to automotive parts to thin films, the material has already been trialled in a number of applications and proven to be equal to or superior to plastics sourced from fossil fuels. Newlight intends to lead its commercialisation efforts with the manufacture of office furniture. But the world's annual plastic use would need to quintuple, with all of it coming from carbon-negative technologies, in order to sequester a gigatonne of carbon per year.

Newlight is one of a number of companies pioneering systems

that allow us to do some of the things that some of the most primitive living things do naturally. The development pathway began in 1953, when American chemists Stanley Miller and Harold Urey stood at a laboratory bench before a flask filled with water vapour, methane, ammonia and hydrogen gas. The scientists hypothesised that the mix was similar to Earth's atmosphere before the origin of life. Two electrodes sparked continuously through the flask, to stimulate the lightning storms that were thought to have occurred early in Earth's history. Within a day, the gases had turned pink, and within a week 15 per cent of the carbon in the mixture had been incorporated into the organic compounds that are the building blocks of life.

The experiment sparked a global media sensation. Had the scientists, in a god-like act, created life in a flask—or at least its precursors? While providing a major breakthrough in our understandings of how life might have begun, the experiment also pointed towards another development that would gather pace more than half a century later. Using water enriched with CO_2, instead of a mix of gases, through which to pass the current, researchers have created long-chain hydrocarbons (the basis of fossil fuels). We are on the very frontiers of science here. Could the creation of oil and other hydrocarbons from CO_2, water and electricity one day become a major route on the third way?

A German company called Sunfire has recently announced that it has discovered a way of creating petrol and other fuels from water and CO_2. The climate benefit of creating petrol this way lies in its potential to replace fossil fuels. The process involves producing steam, and then treating it to remove the oxygen from the H_2O.

Combining the remaining hydrogen with CO_2 leads to the creation of long-chain hydrocarbons. While experimental, Sunfire's work clearly has potential.

Yet another approach is being trialled in Germany. Some years ago the giant engineering company Siemens began working with German universities in an effort to replicate photosynthesis. Were this to succeed, it would bring enormous benefits through the production of many usable materials created from atmospheric CO_2 and water. But photosynthesis is such a complex process that replicating it in the lab was found to be impossible. So, as a first stage, instead of using light energy to split water molecules (as plants do in photosynthesis), the team set about transforming CO_2 into complex hydrocarbons using electricity. The chemistry is complex and as yet poorly understood, but the basic approach involves providing CO_2 with energy. Just what will be produced in commercial quantities using this technology is currently uncertain, but it may be ethylene or various alcohols. A large-scale demonstration facility is due to open in 2015.

Professor Maximillian Fleischer of Siemens says of the project:

> On windy and sunny days, Germany already has more electricity generated from renewable sources than it needs. What it lacks is sufficient energy storage capacity...However, if the electricity were fed into photosynthesis modules, it could be used to produce valuable chemicals. This would help to reduce demand for petroleum and thus cut greenhouse gas emissions. What's more, human beings will have incidentally managed to imitate the most productive chemical process on Earth. The dream of operating biochemical

factories efficiently with sunlight could become a reality.[7]

Currently, all of the chemical technologies outlined require the use of energy—in most cases electricity—and while energy is generated by burning fossil fuels, unless the processes are highly efficient, it's hardly worthwhile using 'dirty electricity' to capture atmospheric carbon. But as Fleischer notes, the wavy baseload provided by wind and solar, with its clean, zero fuel-cost electricity generation, will at some point change all that. The waves in the baseload are caused by the varying generation from wind and solar. But the point is that electricity is always available, and at times is excess to demand. Once wavy baseload comes to dominate Germany's electricity network, using the excess electricity often available to capture and sequester CO_2 will make complete sense.

Who would pay for the costs of capturing carbon? For almost all approaches that deal with CO_2 as waste, the costs will be in the many billions, if not trillions, per year. Under current thinking a carbon price could be used to fund an industry in CO_2 capture and storage, or governments could pay for the service directly out of general tax revenue. The costs are prohibitively high at present. But as the technologies mature, the cost per tonne of CO_2 captured using third-way technologies provides a guideline for the capital that must be raised via taxation and the price of carbon in trading schemes. Additionally, the cost per tonne of CO_2 removed should act as a guide to the price polluters should pay to emit in the first place. Just as the size of a fine given, for example, to the polluter of a lake, should reflect the clean-up cost.

Funding opportunities for environmental remediation (and we can think of third-way technologies as such) are rapidly diversifying.

Perhaps, in years to come some measures will be crowd-funded. Other technologies, such as turning CO_2 into plastics and fuels, may well become the foundation stones of profitable industries, albeit industries that are decades away from profitability at the scale required to draw down gigatonnes of CO_2.

Given their present state, how should we be thinking about third-way technologies in the short term? A recent paper in the journal *Science* argues that we should regard third-way technologies as a valuable series of additional measures that complement our efforts to cut emissions. They should not, however, be used as an excuse for failing to cut emissions from burning fossil fuels. Many challenges remain: the uncertainty regarding costs, side effects and the effectiveness of carbon storage measures, not to mention the need for meaningful accounting. But delaying adoption of these approaches into the political process means that early opportunities may be missed, and research and development may slow, risking the possibility that the technologies will not reach their full potential.[8]

Where might the third way take us by 2050? If we put aside seaweed farming, with its stupendous potential but great difficulties in realisation, the following optimistic scenario is within the bounds of possibility. Forestry and soil carbon might together sequester a gigatonne of carbon per year, and biochar a similar amount. Direct air capture and silicate rocks might capture another gigatonne between them, and carbon-negative cement and carbon-negative plastics another gigatonne. That's four gigatonnes of carbon per year, or around 15 gigatonnes of CO_2—just one quarter of current global emissions and still below the 18 gigatonnes that the combined US academies found we'd need to draw down to reduce atmospheric CO_2 by one part per million per year.[9]

The third way is unlikely to be a cure-all for runaway greenhouse gas emissions. Nor can it ever be considered an excuse for failing to reduce emissions as fast as possible. But, long term, it has the potential to hold back a warming climate to within a range in which human civilisation can thrive. And as such it's a valuable asset that we should be seeking to develop and make the most of right now.

CHAPTER EIGHTEEN

The New Carbon Capture and Storage

Yes, carbon capture and storage, if it can be developed, would help.

ALISTAIR DARLING, CHANCELLOR OF THE EXCHEQUER, 2008

A DECADE ago, geosequestration of carbon, also known as carbon capture and storage (CCS), was looking like an expensive, risky and potentially dead-end technology. That's because most of the investments made have been relatively small and tied to trying to create a future for coal by capturing CO_2 emissions from the smokestacks of power plants, or to injecting CO_2 into oil wells to enhance oil recovery. In other words, CCS was being thought of as a way to prolong the life of fossil fuels. There are inherent limits to these approaches. But, worse, this narrow view of the technology has blinded us to other possibilities.

The role of CCS is particularly relevant to some third-way technologies and approaches. Like battery technology in relation to electric vehicles, large-scale, efficient forms of storage are required to unlock the full potential of many third-way technologies. Before going on to take a new and very different look at CCS, we need to examine progress made thus far.

In newly built power stations, two options exist. Integrated Gasified Combined Cycle (IGCC) methods involve transforming coal into a gas and mixing it with oxygen to fire a gas turbine. The second method involves burning the coal in oxygen-enriched air and recycling the exhaust gases back through the combustion chamber to create an exhaust gas that is mostly CO_2. A large and expensive air separation plant is needed in both methods to provide the massive amounts of oxygen required.[1]

The CO_2 must then be separated from the sulphur, ash, nitrogen and other materials mixed with it and then compressed for transport, usually via new gas pipelines. To store the gas in geological strata, the CO_2 must be compressed even more, until it becomes a liquid. It is then injected into suitable underground rock formations, where it will potentially remain for geological time (thousands or millions of years).[2]

CCS can also be applied directly to natural gas, which sometimes contains large amounts of CO_2. The CO_2 must be separated out, and the purified gas compressed and injected in the earth, as occurs in gas from power plants. There are currently 12 commercial-scale CCS plants operating worldwide. Together they capture and store about 15–20 million tonnes of CO_2 each year, which is only 0.4 per cent of a gigatonne. Eight of these CCS plants inject the captured CO_2 into oil fields as part of the process of extracting the oil.

There is currently only one commercial-scale project capturing emissions from electricity generation (in Canada: it commenced operations in October 2014), though the Kemper County CCS power plant, another such project, is under construction in the United States. Both will use the captured CO_2 for oil recovery.

The Global CCS Institute, headquartered in Melbourne, Australia, expects that by 2020 there will be 21 CCS plants worldwide capturing about 30 million tonnes of CO_2 per year (about 3 per cent of a gigatonne of CO_2).[3]

An inherent problem with using CCS in conjunction with power plants is that 20–25 per cent of the energy produced by the power station is needed to run the capture and storage processes, which makes the electricity that much more expensive to generate. And CCS power plants are also proving more difficult and expensive to build than anticipated. The 582-megawatt Kemper power plant in the US had its opening delayed until May 2015, and is now forecast to cost more than US$5 billion, a substantial increase on the original estimate.[4] At best, the new power station is expected to capture only around 65 per cent of the CO_2 produced (3.5 megatonnes per year). The machinery required to achieve even this is monumental. Photographs of the plant reveal a sprawling monster of pipes and tanks. It looks more like a gargantuan chemical factory than a power plant.

Many people are now concluding that CCS on the Kemper County model is a dead-end. But, recently, some geologists and climatologists have been revisiting CCS. Their ideas of how it might be done and what it might achieve are dramatically different from anything tried previously. And, although fraught with difficulty, some of their pioneering approaches might just achieve the volumes required to make a significant difference to the climate problem.[5] The most promising new thinking is based on finding parts of Earth where conditions are right to store CO_2 in either liquid or solid form. And, surprisingly, those conditions do exist in areas that are not entirely inaccessible.

One of these approaches involves the storage of CO_2 in parts of the ocean crust. Hitherto, those seeking to sequester CO_2 in the Earth's crust on land have faced a daunting problem. As Bahman Tohidi, from the UK's Institute of Petroleum Engineering, and his colleagues explain:

> Because of the subsurface temperature profile of terrestrial storage sites, CO_2 stored in these reservoirs is buoyant. As a result, a portion of the injected CO_2 can escape if the reservoir is not appropriately sealed.[6]

This limits CO_2 storage in continental rocks to geological structures, such as oil-bearing strata, that have impermeable cap-rock or saline aquifers, where natural chemical processes slowly solidify the CO_2. But what if we could use the pressure of the waters of the ocean itself to help keep the gas in liquid form or, even better, to lock it into the rock?

In a series of laboratory experiments, Bahman Tohidi and his collaborators showed that, because of the enormous pressure of the water column, if CO_2 is stored in marine sediments in waters 3000 metres or more deep, it remains in liquid form, although stored only a few hundred metres into the sediment. Over time, natural chemical processes in the water of the ocean sediments convert the liquid CO_2 into a solid—in the form of stable hydrates. The great overlying pressure of the water prevents the CO_2 rising towards the sediment surface, making the storage much more stable than when CO_2 is stored in rocks on land.[7] And when the CO_2 becomes a hydrate, it is locked into the rock permanently.

Although not all regions of the ocean deeper than 3000 metres are suitable for the storage of CO_2, the potential scale of this approach is large. At least a few hundred metres thickness of permeable sediments is required in an area where the topography is not too steep. Steep topography must be avoided because injection of CO_2 there could trigger submarine landslides, which can generate tsunamis. Even with such limitations, researchers note that 'the total CO_2 storage capacity within the 200-mile economic zone of the US coastline is enormous, capable of storing thousands of years of current US CO_2 emissions'.[8]

While the research is at an early stage, it does suggest that the option should be examined further, and even prioritised. Together with a steep reduction in emissions and seaweed farming, CO_2 storage in deep water marine sediments might just be planet-saving.

Another intriguing proposal for the geosequestration of CO_2 was recently published by Professor Ernie Agee and his colleagues at Purdue University in Indiana. It concerns the potential capture and storage of the gas in the Antarctic ice cap. The research was triggered by the observation that Mars has polar ice caps composed of frozen CO_2, and the scientists wondered if Earth's ice caps might be capable of storing frozen CO_2 as well.

It turns out that conditions over the Antarctic ice cap are so severe that the storage of solid CO_2 (dry ice) might be possible. At sea level, CO_2 freezes at –78.5°C. The Antarctic ice cap has an average elevation of about 2500 metres, and temperatures of –89.2°C have been recorded at Vostok Station on the ice cap. This is close to the temperature required for CO_2 to freeze and begin to accumulate as snow. At higher temperatures, CO_2 freezes over the Arctic ice cap, but sublimates (changes back to a gas) as quickly as it freezes out. The average temperature over the interior of the Antarctic ice cap is –57°C, so in most weather conditions only about 30°C of cooling would be required to cause CO_2 to fall out of the air.

Professor Agee and his colleagues propose building a series of 100 × 100 × 100 metre refrigeration chambers high on the Antarctic ice cap. Air cooled with liquid nitrogen to below CO_2's feezing point, would cause the precipitation of about 40 centimetres of CO_2 snow per day, leaving all other components of the air in gaseous form. The CO_2 snow could be stored in pits in the Antarctic ice, and covered with ice and snow to prevent its loss through sublimation on exposure to the slightly warmer air. The researchers estimate that only sixteen 1200-megawatt wind plants (less wind power than currently exists in Germany) could provide all of the energy required to drive 446 such cooling chambers. And that would be enough to capture and store one gigatonne of CO_2 per year.[9]

Antarctica is a windy place, and wind power is already in use at research stations on the continent. Moreover, an existing global treaty—the Antarctic Treaty—provides a framework for scientific co-operation and international governance. Because the proposed refrigeration chambers are modular, it's possible to build a trial plant in order to investigate this proposal further.

One important objection to the proposal is that Antarctica is Earth's last continental-scale wilderness. Many people would be reluctant to see large refrigeration cubes and wind farms scattered over its surface. But it seems to me that these objections share much with those of people who object to wind farms in the countryside on the basis they don't like looking at them. As the climate problem grows, such objections will surely need to be weighed against the growing climate impacts.

What other possible downsides might there be to storing CO_2 ice in Antarctica? One potential problem is that the concentration of CO_2 in the air over Antarctica might become greatly reduced, and this could affect the surrounding ocean, or indeed the southern hemisphere as a whole. And it's possible that the very cool air might affect global atmospheric circulation. Both possibilities are easily investigated with climate modelling, and this should be done. Indeed even simple observations of what happens when air temperatures drop below –78.5°C naturally would be highly useful in understanding the potential effects. In any case, CO_2 levels could be kept within the historical range experienced over Antarctica by regulating the speed of the drawdown, and since the atmosphere mixes readily, the local CO_2 depletion would not last long after the plant throttled back.

There is also the risk of CO_2 escape, if the ice cap were to warm. The conditions under which such a release might occur require investigation, though they are extremely unlikely even on the thousand-year timescale. If they did, Earth would be facing a full-blown climate crisis in any case. Conversely, if we take a long view and examine a future Earth threatened with an ice age, the trapped CO_2 could be exposed and allowed to warm the atmosphere.

The cost of building the proposed infrastructure in Antarctica is very difficult to estimate, but it is likely to be huge. Back of the envelope cost estimates of up to a trillion dollars might be conservative.[10] But technologies and cost structures change, and with the project unlikely to be seriously considered before 2050, both costs and funding models may then be very different. If the technology, the functionality and the economics of the proposal became increasingly robust, it's not difficult to imagine such a project as the first globally crowd-funded project to save humanity's future.

While these possible applications of geological storage of CO_2 may seem distant and even dangerous to some, we must calibrate that against the dangers of living with 450 ppm or more of CO_2 in the atmosphere for a century or longer. Whether that danger becomes a reality will be determined by the success of our efforts in reducing carbon pollution. However, we are on a worst-case emissions trajectory at present, so if nothing changes, in coming decades the idea of storing CO_2 as snow at the South Pole, or deep in the ocean's sediments, may not look so risky after all.

CHAPTER NINETEEN

The 2030 Challenge

Just because you are struggling does not mean
you are failing.

ANONYMOUS

THE extent that emissions are reduced by 2030 will be strongly influenced by the outcome of the Paris climate summit. As noted earlier, the aim of giving humanity a 50–50 chance of keeping temperatures from rising by less than 2°C above the pre-industrial average is not ambitious. Even if it succeeds and we are lucky and temperature rises stay within 2°C, the world may still suffer serious damage. If it succeeds and we are unlucky, we could be faced with catastrophic climate change. That is why I argue so emphatically that humanity must begin working on third-way technologies now, so that they will be mature enough if, and when, we need them.

The accounting firm PricewaterhouseCoopers notes that our current trajectory is not conducive to an emissions cut. Every year that passes is bringing us closer to a human-caused climate disaster, the accountants say. The unmistakable trend they see is that the world's major economies are increasingly failing to do what's needed to limit global warming to 2°C above pre-industrial levels.[1] Another gloomy message was recently delivered by Fatih Birol, chief economist with the International Energy Agency. On 3 June 2014, in an interview with *Energy Post*, he highlighted the fact that progress in limiting coal and moving to renewables had slowed between 2012 and 2013.[2] His boss, Maria van der Hoeven, repeated the message: 'Growing use of coal globally is overshadowing progress in renewable energy deployment.'[3]

Birol emphasised that the recent gradual shift away from fossil fuels and towards renewables wasn't enough.[4] A dramatic change of course is required if the world is to avoid a climate-change disaster. And, of course, dramatic shifts require decisive government action—to limit coal use and encourage new technologies by putting an effective price on carbon. Otherwise cheap, dirty coal, with its vast already-built and paid-off infrastructure, will prevail.

A study released by Oxford University's Stranded Assets Program quantifies just how large the task of reducing coal use is.[5] About 1617 gigawatts of coal-fired power capacity currently provides about 40 per cent of the world's electricity. Seventy-five per cent of coal's share is generated by old, inefficient, subcritical coal plants. If we are to have even a 50–50 chance of keeping warming to within 2°C, 290 gigawatts of this capacity will need to be shut down by 2020. That means shutting down nearly 300 really large coal-fired power plants, or a much greater number of smaller ones, over the next five years. Given the electricity deficit faced by many developing countries (which is where many of the most inefficient coal plants are), this is a huge task. Even wealthy countries such as Australia have proved unwilling to shut their filthy, antiquated coal plants.

So what can we expect of the various nations as they clamp down on emissions? There are, surprisingly, grounds for hope. The spectacular success of the US in reducing its emissions back to 1990 levels, China's determination to assume leadership in the clean energy area and Europe's decades-long leadership in reducing emissions are causes for optimism. But the torpedoing of the Copenhagen meeting by those wishing to frustrate action is still fresh for many. They fear that governments opposing a price on carbon and other actions—such as the Abbott government in Australia and Canada's Harper government—will act to derail a global agreement, perhaps by forging a coalition with other nations that are resolutely opposed, half-hearted or undecided, such as Russia, India, Saudi Arabia and Turkey. Despite the clear lead given by the major polluters, the consensual requirements of global climate action mean that such a 'coalition of the unwilling' could easily frustrate or slow global action.

Regardless of these concerns, in early 2015 hopes for the Paris meeting were high. The prospects for success have received a series of boosts, the first and arguably most important being the Obama administration's announcement that the US will move on emissions in ways that are difficult for Congress to block. On 2 June 2014, a rule to restrict CO_2 emissions from existing power plants was announced. It allows states to reduce their carbon emissions in the ways that best suit their individual electricity resources portfolios. Increasing their use of clean energy and improving energy efficiency are options, along with or instead of the installation of pollution controls at the smoke stack, a requirement that has typically been blocked by Congress. The proposal has met with surprisingly little criticism from the coal industry, even though it will reduce CO_2 emissions by 30 per cent from 2005 levels by 2030.[6]

Obama's Clean Power Plan 'helps the administration's negotiating position to the extent that they can now credibly say they have a plan in place to meet the US's short-term emission reduction commitment', says Evan Juska, who heads US policy at the Climate Group, an international non-profit organisation.[7] Another move by the Obama administration is being mooted. After flattening out around 2005, the concentration of atmospheric methane, a powerful greenhouse gas, has begun to increase again. One contributor is the expanding gas industry, particularly due to fracking in the US, where emissions have recently been estimated to be around 50 per cent greater than previously thought.[8] Methane from the oil and gas industry is the last major unregulated emissions sector in the US. One possibility is that the Obama administration will move to reduce methane emissions from the sector by 45 per cent of 2012 levels by 2025.[9]

In the early hours of 24 October 2014, the EU announced its own plan to reduce emissions: Herman Van Rompuy, president of the European Council, tweeted: 'Deal! At least 40% emissions cut by 2030. World's most ambitious, cost-effective, fair #EU2030 climate energy policy agreed.'[10] Such deep cuts by the EU were, according to commentators, at the most ambitious end of the spectrum of possibilities, and while not sufficient to avoid warming of more than 2°C, they are in line with the kind of action hoped for at the Paris meeting.

At the end of the 2014 APEC meeting in November in Beijing, President Zi and President Obama delivered more important news:

a bilateral deal to address climate change. Negotiations had been going on for nine months, but had, astonishingly, remained secret. The pledges announced push out obligations to 2025 for the US, and 2030 for China. But, for the US they deepen ambition from 17 per cent emissions cuts on 2005 levels (4 per cent on 1990 levels), to 26–28 per cent on 2005 levels (14–15 per cent on 1990 levels). The US is aiming for an 83 per cent reduction by 2050 on 2005 levels (80 per cent on 1990). If the US implements policies that achieve its 2025 pledge, this would begin to place the nation close to, but still on the high side of, a pathway consistent with limiting warming to within 2°C.[11]

China committed, for the first time, to an emissions peak 'before, or no later than, 2030'. This will most likely leave China far from contributing its fair share to a world no more than 2°C warmer. Carbon Tracker says of the announcement:

> What is not clear, but is critical, is the peak level of China's emissions. In the assessment here Chinese peak emissions would lie far above a 2°C consistent emission pathway. We see however significant potential for improvement.

The Chinese pledge to increase its clean energy target may be more significant. It commits the country to building more clean-energy infrastructure over the next 15 years than it has built in coal power in all of China's history. Although the allowable technologies include hydro and nuclear, the massive amounts of wind and solar required to meet the target will revolutionise both industries, pushing costs down further and so making the technologies more competitive.

According to Niklas Höhne of the NewClimate Institute, the

effect of the China–US deal is that 'both move closer to a pathway that is compatible with 2°C but would need to increase ambition'.[12] At the same time, the deal revitalises the climate negotiations, adding to the likelihood of an acceptable outcome at the Paris summit in 2015.

A few weeks after the China–US deal was announced, the climate meeting in Lima, Peru, after running two days over schedule, came up with an agreement that Sam Smith, chief of climate policy at WWF, described as going from 'weak, to weaker, to weakest'.[13] There will be no requirement for nations to provide quantifiable information detailing how they will achieve their targets. But, on the positive side, every country on Earth has pledged to produce a plan of action to tackle climate change, and political leaders saw it as a vital step towards concluding a deal in Paris.

It's unclear whether the agreement to be concluded in Paris will be a treaty, or a Copenhagen Accord–like deal. The US has the particular problem that its Republican-controlled legislature is almost certain to refuse ratification of a global climate treaty. This dilemma has caused President Obama to shy away from the mechanism of a treaty and, instead, to seek to build a different kind of global agreement—one based on the Copenhagen Accord. The Europeans and many developing nations would prefer a treaty.

If President Obama can broker a strong but not legally binding international agreement in Paris, it would not require ratification by the Republican-controlled Congress. British economist Lord Stern has been persuaded by Obama's position, recently stating: 'that way you will get stronger ambition as countries won't be tempted to be hesitant about some type of international sanction'.[14] Yet some fear that a non-treaty agreement will not carry the force of law. To be effective, such an agreement would rely on a kind of public scorekeeping, what some describe as 'name and shame', to encourage countries to meet their carbon commitments. 'It's probably not as effective as a legally binding treaty would be, but it can encourage countries to do more than they otherwise would,' Evan Juska of the Climate Group says.[15]

It's not certain that any agreement made in Paris will include pledges for emissions cuts out to 2050. Here the split is between the developed world (who would like to include such pledges) and developing nations, who fear that 2050 pledges may lock them out of using fossil fuels before their economies can afford to do so.

Just how far the Paris meeting might take us towards a stable climate can be inferred from the declared positions of China (categorised as part of the developing world) and the US (a developed nation). They are jointly responsible for 44 per cent of global emissions from fossil fuels. As we've seen, although their recent announcements are considerable, their combined actions are unlikely to see us on a safe climate trajectory. It's to be hoped that the decisive action of the EU will act as a benchmark and challenge for the nations responsible for the remaining 55 per cent of emissions. But it seems unlikely that Canada and Australia, both major emitters, will commit to aggressive emissions cuts. If most developed countries follow the European lead, over the longer term the actions of the G77—the 77 nations classified as developing (which includes China)—will play a big role in determining whether we overshoot the summit's target by a lot, or a little.

Can the pledges made thus far be honoured? Can China really build more renewable energy infrastructure in the next 15 years than it has built of coal-powered plants in its entire history? A price on carbon is surely a prerequisite to success, and thankfully China's plans for a nationwide emissions-trading scheme are well advanced. The government is close to finalising the scheme's rules, and has announced that it will begin in mid-2016.[16] The proposed price of carbon in the Chinese national carbon market is not yet known, but prices on the various regional markets range from US$3.60 per tonne of CO_2 emitted in Hubei to $20.00 in Shenzhen. It is reasonable to assume that the national price will fall somewhere in-between.[17] Despite its cautious pledges, China's dramatic reduction in coal use, together with its support of renewables and electric vehicles, puts it in a very strong position to argue at the Paris meeting that it is leading by example, and that others should follow. Other Asian nations are indeed on the same path: South Korea launched its national emissions-trading scheme on 1 January 2015, and Indonesia, Thailand and Vietnam are drawing up plans for carbon-trading schemes of their own.[18]

Emerging data from China provides some very welcome news. It suggests that the nation is about to, or already has, hit peak coal use. If this is the case, provided its incipient emissions-trading scheme has a sufficiently high price, it is likely that China has already done most of the work required to get emissions from electricity production under control. But when it comes to transport, China's challenge remains formidable. There are currently about 120 million passenger cars on the road in China, and about 15 million new cars added each year. The US has four vehicles for every five people, and there is no doubt that China is heading in the same direction. A conservative estimate is that there will be 300 million passenger vehicles on Chinese roads by 2030—one car for every four people. Even if those cars are all running at efficiency levels that exist in Europe today, despite China's great work in curtailing coal use, they will keep the nation's emissions where they are now.

In essence, this is why China is so reticent about committing to an emissions peak prior to 2030. Electric vehicles could, of course, change all that. But China's largest EV manufacturer built just 2000 cars in 2013. It's hard to imagine that it could be building 15 million a year between now and 2030—or indeed that China could import even a tiny proportion of that number. Just how transport plays out in China will have a very big impact on our common future.

An interesting assessment of the opportunities for more ambitious emissions cuts by both the US and China was recently published by the Climate Action Tracker, an NGO that tracks emissions commitments from all countries. It observed that both the US and China have instances of world's best practice in reducing carbon emissions, but only in certain sectors. In each category examined in the report (electricity, buildings, transport and industry), Climate Action Tracker found that if China and the US each matched the other in their best-performing sectors, they would reduce their emissions to levels below current projections for 2020 by 170 megatonnes in China and 220 megatonnes in the US, and for 2030 by 1100 megatonnes in China and 1100 megatonnes in the US.

Even better results could be produced if China and the US were to adopt global best practice in each sector, with the US achieving a reduction of 18 per cent on 2005 levels by 2020 and a 32 per cent drop by 2030, though this figure doesn't take into account land use and forestry. China's emissions would peak by the early 2020s and then begin to decrease, but, combined, their emissions would be 2.8 gigatonnes of CO_2 below current policy projections and account for 23 per cent of the emissions reductions needed to keep the planet within 2°C of warming.[19] The study demonstrates how much existing technology can do to limit climate change. We clearly have the tools needed to avoid more than 2°C of warming. But will we use them?

Among the countries most closely watched in the lead-up to Paris is India. Inevitably, it will host the next big boom in energy demand. More than 300 million Indians still lack access to electricity and, as of 2012, India was still 59 per cent dependent on coal for electricity generation.[20] So the big question is: will Indian demand for electricity blow the global carbon budget?

India's energy infrastructure is creaking. Electricity theft accounts for losses from the network of around 23 per cent. In Old Delhi, television remote controls are modified so that they can be used to switch electricity meters on and off, and meter readers have been threatened when they try to install new tamper-proof meters. In slums across the country one can see the illegal, makeshift wires tapping into the network. It's not surprising that investment in electricity infrastructure has been low. As a result, in 2013 India suffered the largest blackouts experienced anywhere on Earth: 660 million people were left without electricity.

To date, wind has been Indian's great renewable success story, at least as a manufacturing industry. Established in the 1990s, Suslon is India's global wind power company. India is currently the world's fifth-largest wind producer, after the US, China, Germany and Spain.[21] Domestically, though, wind is coming off a low base, with a total installed capacity, in January 2014, of just 20 gigawatts, and growth of around 5 per cent per year. India is now exploring offshore wind, which could greatly expand opportunities.

Solar PV is slowly beginning to grow as well. The most popular dowry gift in India today is a solar-powered lighting kit. Despite the wide public appeal of solar, in the absence of aggressive government action, uptake has been slow. The government's National Solar Mission and state-based solar incentives together added less than one gigawatt of solar power capacity between April 2013 and March 2014, and as of April 2014, India had only 2.6 gigawatts of installed solar capacity and no manufacture of solar panels.[22] Changes, however, are underway. The Modi government has increased India's national solar target fivefold to 100 gigawatts—the same level China is aiming for by 2020. As a result, India's solar market is now the fourth largest in the world, up from eighth in 2013.[23]

Encouragingly, on a recent visit to India, President Obama pledged to help finance India's Solar Mission and hopes for support from Modi at the Paris climate summit in exchange.[24] At a joint press conference with Obama, the Indian prime minister said, 'India is ready to expand its use of renewable energy as a way to reduce greenhouse gas pollution.'[25] Despite Obama's efforts, affordable financing remains a major barrier to solar's rapid expansion. Some funds will be raised by government. The 2015 budget includes provision for a doubling of India's coal tax to US$3.40 per tonne, the proceeds of which are to be invested in clean energy.

The scale of India's planned energy revolution was laid out in November 2014 when India's new energy minister, Piyush Goyal, presented a five-year plan. Among the problems it will tackle are a lack of diversity in generation, providing electricity to the 300 million Indians who lack it and the need to combat growing air pollution.[26] The plan includes a 50 billion rupees investment in transmission upgrades, and 100 billion rupees for new renewable energy installations. Coal India is being directed to invest 1.2 billion in solar projects, bringing to bear part of the US$10 billion in cash assets sitting on its books. With these investments, India plans to build the most advanced power grid anywhere. An improved grid will bring even greater efficiencies (though India's perpetual energy scarcity preconditions its industry to efficiency), as well as allowing for greater penetration of renewable energy.

As Goyal acknowledged, off-grid solutions are also required. Narendra Taneja, convenor of the Indian ruling party's energy division, says that by 2019 every home in India will have enough solar power to run two light bulbs, a solar cooker and a television.[27] As chief minister in Gujarat, Narendra Modi pioneered the uptake of large-scale solar in that state, so he has a good track record with ambitious clean tech schemes. If sufficient financing is secured, India's expansion into solar will mark a major shift that may well be looked back on as one of the most significant initiatives in the lead-up to the Paris meeting.

After India, Africa is the last continental-scale 'greenfields' site for electricity generation. Six hundred million Africans do not have access to electricity, but things are changing very quickly, and renewables are taking up the slack. Hydro power is an important power source throughout equatorial Africa, and plans are underway to massively increase its capacity. But other forms of renewable energy are also taking off. In mid-2013, Africa had just one gigawatt of installed wind capacity. But that figure is set to grow tenfold according to the African Development Bank.[28] Indeed, in 2014 Africa added more renewables—mostly wind and solar—than it had built in the preceeding 14 years. Solar technologies are particularly suited to African conditions.

Today, coal provides 94 per cent of South Africa's energy needs. Elsewhere in Africa, grid-based electricity is absent or is largely supplied from fossil fuels. But according to one analysis, a 60-kilometre-square area of concentrator solar power stations (which use reflective surfaces to concentrate sunlight) could provide the Middle East and North Africa's total electricity requirements.[29] With a projected population of four billion by 2100, Africa must take a renewables pathway to development if we are to remain below the 2°C threshold.

Elsewhere in the developing world, a diverse mix of renewables is being installed, with one of the surprises being geothermal energy. Geothermal energy takes many forms, but one of the most promising in the developing world is the large power stations, under construction, that use heat from volcanic rocks to generate electricity. At the Climate Project Asia Pacific Summit in 2011, Al Gore stated that Indonesia could become a superpower in electricity production using geothermal energy.[30] Indonesia is soon to start building the world's largest geothermal plant. The US$1.17 billion

Sarulla project in north Sumatra will provide 320 megawatts of emissions-free electricity, and is just the first step in what may be a major energy transition. India has also recently announced plans for the country's first geothermal power facility, in Chhattisgarh state in central India.[31] Britain is currently pioneering wave and tidal energy technologies, and these may join geothermal energy in powering developing countries after 2020.

The pace at which wind and solar have penetrated developing markets has surprised everyone. When these are combined with the potential of geothermal, wave and tidal power, there is now real hope that much of the developing world can circumvent the energy path taken by the developed world and China, which involves heavy use of polluting energy generation. But it is not yet certain that this will happen. Much of the developing world is in the earliest days of an energy revolution; the outcomes at Paris will have a huge impact on their future development pathways.

One of the most encouraging signs that a clean energy pathway will open globally was the announcement, by the International Energy Agency in March 2015, that the growth in global emissions from burning fossil fuels to generate energy had 'stalled' in 2014 at 32.3 gigatonnes—the same volume as in the previous year.[32] The preliminary announcement did not identify with any certainty the cause or causes of the stall, but commentators noted that China's declining coal use could be a factor, as could the weather and the relative pricing of fossil fuels.[33] Nonetheless, IEA chief economist Fatih Birol described the news as 'a very welcome surprise and a significant one'. All previous pauses or declines in emissions have occurred during economic downturns. But this was not the case in

2014. 'For the first time, greenhouse gas emissions are decoupling from economic growth,' Birol said. Thus, it's possible that the world as a whole is decoupling economic growth from emissions growth. But we cannot know whether we've passed 'peak emissions' for greenhouse gases until we see clear evidence of a years-long decline.

CHAPTER TWENTY

Deadline 2050

> Yes, we may find technological solutions that propel us into a new golden age of robots, collective intelligence, and an economy built around 'the creative class'. But it's at least as probable that as we fail to find those solutions quickly enough, the world falls into apathy, disbelief in science and progress, and after a melancholy decline, a new dark age.
>
> NEIL DEGRASSE TYSON, *COSMOS*

OF COURSE it is good to plan for distant targets. But in climate negotiations thus far the idea of planning for 2050 is proving more of an impediment than a benefit. Many developed countries want a zero target for carbon emissions by that date, while developing countries want to keep open the option of having some fossil fuel use beyond that date, in case they cannot make the transition to clean energy fast enough. Despite very lengthy discussions no agreement has been reached. Numerous nations, states and cities have, however, set 2050 emissions reduction targets.

The pledges are typically of reductions of at least 80 per cent on 1990 levels, with some pledging to eliminate carbon emissions entirely. Understandably, the pledges are short on detail, but this does not render them meaningless. Rather, they are useful political instruments for indicating the earnest intentions of the parties involved.

The problems nations face in detailing plans of action out to 2050 are formidable. To gain a sense of the difficulties involved, imagine you are living in 1915, and trying to set targets for technological change, politics and the environment as they might exist by 1950. In an era of canvas biplanes, horse-drawn wagons and empires, it would be near impossible to foresee jet aircraft, atomic power, the spread of Communism or the ubiquity of automobiles. The rate of change in the twenty-first century is already greater than that of the century preceding it. Many believe the world of 2050 is genuinely unimaginable from today's perspective, and that therefore laying out detailed roadmaps is pointless.

What should we be aiming for by 2050? In 2007, a group of researchers set out to assess the long-term climate change implications of national emissions reduction targets for 2050. The University of Victoria's Andrew Weaver and his team concluded that:

> Even when emissions are stabilised at 90 per cent below present levels at 2050, [the] 2.0°C threshold is eventually broken. Our results suggest that if a 2.0°C warming is to be avoided, direct CO_2 capture from the air, together with subsequent sequestration, would eventually have to be introduced in addition to sustained 90 per cent global carbon emissions reductions by 2050.[1]

It's important to understand that when Weaver and his team talk of the 2°C threshold, they are talking about a 50/50 chance of staying below that threshold. In order to give us a 75 per cent chance, we'll need to decarbonise earlier and faster. That's why so many believe that for humanity to have the best chance of avoiding a world warmer than 2°C above the pre-industrial level, the burning of coal, oil and gas without carbon capture has to be a distant memory by 2050. And even then we will also probably need third-way technologies to reduce the atmospheric burden of CO_2. So, over the next few decades some of the largest corporations that exist today—among them Exxon Mobil, Anglo American and BG—must either lose all value and so cease to exist or change their business model completely. And in their place other corporations must emerge, investing trillions of dollars on clean energy.

A recent UN study outlines how global emissions might be cut deep and fast. The Deep Decarbonisation Pathways Project (DDPP), headed by the United Nations secretary-general, Ban Ki Moon, shows how 15 countries (Australia, Brazil, Canada, China, France, Germany, India, Indonesia, Japan, Mexico, Russia, South Africa, South Korea, the UK and the US), which together account for 70 per cent of the world's greenhouse gas emissions, could together cut their emissions in half while tripling economic output. So, the roadmap to achieve steep reductions exists. But, dismayingly, the DDPP found that very few countries have looked seriously at how they might do their bit to ensure that we stay within the 2°C limit.[2]

The International Energy Agency has examined the problem from a different perspective, asking how much we would need to ramp up deployment of renewables if we were to stay within 2°C. Together, solar, solar thermal and wind need to be contributing half of all electricity generation by 2015. Solar's share (16 per cent) would require deployment of 124 gigawatts, at a cost of US$225 billion, annually. That's two or three times the industry's current level of growth. Wind's share (15–18 per cent) requires its annual rate of deployment to rise from 45 gigawatts in 2012 to 65 gigawatts by 2020, 90 gigawatts by 2030 and 104 gigawatts by 2050. Solar thermal's contribution (11 per cent) will need to grow from almost nothing today.

The IEA's 2014 report on future global energy use sees a very different future. It projects that renewables will be contributing only 33 per cent of global energy by 2040 and that demand for all fossil fuels will grow until that time.[3] Depressingly, this all adds up to a world which has lost the opportunity to keep warming within 2°C, but instead is committed to a world 3°C warmer, or even more. As Bill McKibben so ably puts it (referring to the fossil fuel industries): 'Their vast piles of money have so far weighed more in the political balance than the vast piles of data accumulated by the scientists.'[4]

For all the opinions and uncertainties with which we approach this world just 35 years away, there are a few enduring truths that can be relied upon. One is that innovation is driving economies as never before. But while innovation is transforming the economics of renewable energy, innovation is not always green. Fracking, tar-sands mining and the petrochemical industry are all benefitting from innovation. We can only hope that manufacture of solar PV and wind turbines, with their astonishing capacity to get costs out of the production processes, will benefit more from innovation than the fossil-fuel industries do.

Another truth is that the volume of fossil fuels within our grasp is growing. As recently as a decade ago climate scientist James Hansen felt that while humanity had to limit the burning of coal, it could exploit all the known sources of oil and gas without breaching the 2°C threshold.[5] Fracking and other unconventional ways of mining have so expanded accessible reserves of oil and gas that they have demolished that comforting idea. We now know that long before 2050 we must have eliminated the burning of all fossil fuels, if we are to stave off a climate crisis.

CHAPTER TWENTY-ONE

The Growing Power of the Individual

The mind is the limit.

ARNOLD SCHWARZENEGGER

SOMETIMES you don't discover your strength until your darkest hour. That was certainly true for me in 2013 when I lost my job as Australia's first Climate Commissioner. For almost three years I'd been tasked with providing Australians with access to the latest developments in climate science, economic thinking about climate change and political action worldwide. Working with some of the nation's top scientists, economists, business leaders and bureaucrats, I also met thousands of Australians face to face and helped answer their questions. In addition, the commission produced 18 climate reports and set up a highly informative and accessible website. The first actions of the newly elected Abbott government were to sack me, abolish the commission and take down the website.

I have rarely felt so helpless or frustrated. I knew that the commission's work was vital if Australians were to understand the huge challenges ahead, but I could see no way of continuing our work, until Amanda McKenzie, one of my colleagues at the commission, suggested that we turn to the Australian people for help. Together the commissioners decided to take up the challenge, and, just five days after being sacked, we launched a crowd-funding campaign. A week later, ordinary Australians had contributed almost a million dollars, and we were on the way to setting up the Australian Climate Council. Its objectives are identical to those of the Climate Commission but, freed of government shackles and with a budget of A$1.75 million per year, it is a far more active and effective organisation.

The establishment of the Climate Council so soon after the abolition of the Climate Commission was a great victory for Australians who care about climate change. The organisation now acts as a major reference point for climate action nationally, and we have helped to change the way many people think about the issue. In 2015 more Australians understand that climate change is a problem than did in 2013, and support for action to address climate change is growing.

Just a few years ago none of this would have been possible. Crowd-funding platforms are relatively new, as are the social media the Climate Council uses to communicate. The take-home message for me is that individuals now are immeasurably more powerful in the battle against climate change than they were a decade ago. This is, in part, because social media can create communities of interest; and communities, being composed of customers and voters, are capable of altering the policies of institutions and corporations. Sheer creativity also means that the number of options for activism has proliferated. Following are just a few possibilities that I hope will inspire you.

If you are one of the 2.6 million Australians who have solar panels on your roof, and you wish to defend renewable energy, you can join Solar Citizens. This group, composed mainly of ordinary Australian retirees or people with mortgages (both of whom are often on fixed budgets and watch electricity prices closely), is becoming a formidable force in politics. When the Australian government recently threatened the nation's renewable energy target, Solar Citizens began organising action in the 10 electorates where government members hold their seats with slim majorities. Their work has been an important influence on the government moderating its stance.

For young people, there's a variety of national action groups united under the global Youth Climate Movement. In North America the Energy Action Coalition runs a series of events and actions, including 'Fossil Fools Day' which is held on 1 April and is marked by meetings and activities on campuses across the US aimed at reducing dependence on fossil fuels. Their Power Shift conferences and Campus Climate Challenge programs agitate for more renewable energy, while their Power Vote campaign encourages younger citizens to vote for those advocating action on climate change. In Australia, there's the Australian Youth Climate Coalition (AYCC). Membership is restricted to those under 30, and social media is its preferred means of action. The AYCC's Powershop campaign is typical of its many works. The campaigners use social media to organise people to shop at a particular store at a particular time—say, a supermarket at a weekend—after arranging that the profits from the weekend sales be put towards renewable energy or energy efficiency. Many businesses now owe their solar panels or energy efficiency measures to Powerup. And the program has become so popular with businesses that many have requested that the events be repeated.

Another highly effective project carried out by AYCC involved personal visits to the branch managers of Westpac, one of Australia's biggest banks. Two hundred of the bank's 300 branches had been visited by mid-2014. The AYCC campaigners explained to management that they were concerned that climate change fuelled by new coalmines in Australia's Galilee Basin would affect their futures, and asked the bank not to invest in such projects. News of the visits soon reached headquarters, and Westpac is now fully aware of the extent of community opposition to the mines. Looking back over the past decade, I'm delighted at how effective activism by young people has been. When I wrote *The Weather Makers*, I remember how helpless many students and young people felt in the face of a climate crisis created by an older generation. If their elders had been half as effective as they are, we would have the climate problem under control by now.

Many options that were hard for people to adopt a decade ago are now easy. In 2015, for example, installing solar panels is common sense on economic as well as environmental grounds, as is buying a fuel-efficient vehicle. Due to improved infrastructure, options like riding a bike or walking are also easier than ever. In many cities public transport options are increasing, with light rail networks in particular expanding. The various disinvestment campaigns now being waged on university campuses and in superannuation funds offer yet another course of action not available or not effective a decade ago. And, given the drop in the value of many fossil-fuel companies, divestment is looking like wise investment.

The entire electricity sector—from generation to transmission and retail—is being transformed by community action. Auckland-based Vector provides one example of how this is being done. The New Zealand company owns transmission assets in electricity, gas and communications (fibre-optic cables). It is 75.1 per cent owned by a trust composed of the company's customers, who received an annual dividend of NZ$335 in 2014. Vector is now selling battery storage to its customers, so that its investment in expensive electricity transmission assets can be minimised. In Germany, communities are now campaigning to have ownership of transmission assets (poles and wires) transferred to local governments or co-operatives.

The selling of electricity is also transforming in Germany, where some electricity retailers are already community-owned. In Australia, plans to develop a community-owned electricity retailer in northern New South Wales are also well advanced, and many other examples exist in the US and Europe. The story is the same in electricity generation. The wind industry arose in Denmark in the 1970s with community-owned wind turbines. Thanks to innovative financing models, community-owned electricity generation is spreading rapidly in North America, Europe and Australia, with community-owned solar experiencing a boom even larger than that seen in wind. There's a touch of 'back to the future' about this. Just a few decades ago electricity utilities were often owned by local or state governments. Then neoliberal economics saw them sold to corporations. Now, courtesy of new technologies, power is returning to the people.

If litigation appeals as a means of community activism, it's worth knowing that the grounds on which governments are being taken to court for neglecting climate change are growing. In January 2015, the Mackay Conservation Group, a small group of environmentalists based in central Queensland, challenged the Australian government's approval of the Carmichael coalmine in the Galilee Basin. At the heart of the challenge were the impacts of the CO_2 on Australia's Great Barrier Reef. The litigants claimed that the minister for the environment, Greg Hunt, had erred in ruling that the greenhouse gas emissions that would be created when the coal was burned were not relevant to his assessment. 'That's what we say is a major breach of Australia's environment laws,' said the group's spokeswoman, Ellen Roberts.[1] It's the first time that a proposed coalmine has been challenged on such grounds in Australia, and the Environment Defender's Office, a community legal centre specialising in public-interest environmental matters, is assisting the group. Samantha Hepburn, professor of law at Deakin University, says, 'This has the potential to be a landmark decision and the Federal Court will be examining the role of the national environment legislation in terms of, ultimately, climate change and global warming.'[2]

Another potentially far-reaching legal case was recently instigated in the US by the group Atmospheric Trust Legal Actions, an entity created by young people, including law students, at Oregon State University. They took on the Oregon state government, demanding that it protect the climate. Their case was dismissed by a lower court, but in January 2014 three Oregon Court of Appeals judges visited Oregon State University to hear the appeal in its classrooms, giving law students the opportunity to witness an appeals court hearing on their own campus.

The students' argument springs from Professor Mary Christina Wood's book *Nature's Trust*. Published in 2013, *Nature's Trust* argues that citizens have a right to live and flourish. Therefore, a government elected by the people has a duty to protect the natural systems required for their survival: forests, wildlife, soil, water and air.[3]

Nine lawsuits or petitions based on Wood's innovative legal theory are currently making their way through US state and federal courts as well as courts overseas. Her framework calls upon the public trust doctrine—which holds that certain resources are owned by and available to all citizens equally—to enforce the constitutional right to a livable environment. It includes the atmosphere as an asset in that trust, and it calls government to 'restorative duty', which means not just preventing future damage, but repairing past harms that scientists now identify as threatening to current and future generations.

On 11 June 2014, the US Court of Appeals reversed a lower court's dismissal of the Oregon students' case. No matter what happens in these lawsuits, Professor Wood argues that legal action:

> is not going away for two reasons. First, the climate crisis is intensifying, and courts are going to change their view of their role as more heat waves strike and the legislature sits idle. And second, the public trust doctrine isn't going away. It's been around since Roman times. And it is really too deep for any one opinion—even a Supreme Court opinion—to wipe out.[4]

Envoi

I HAVE mixed feelings about our future. Emissions have grown too fast, and their impacts on the climate system have been severe. And I've had my hopes dashed once before, in Copenhagen in 2009, making it hard to place my remaining reserves of hope in the Paris climate meeting. But I also feel that things have changed fundamentally, and for the better, in the last few years. The threats are accelerating beyond expectation, but so are the opportunities. There is real hope in the IEA findings that global emissions growth has stalled even as the economy has grown. And the clean energy revolution, with its promise of clean power and electric vehicles, now seems unstoppable. Moreover, the arguments of the deniers are now transparently wrong, and outdated. The thought that they might deny the world a last chance for a better climate future is perverse. Even grotesque.

I grew up in an age of technological optimism. We thought that each decade would be better than the last. I thought that this was the way it would always be. But speaking to young people today, I realise that they have a very different view. They fear that their lives will be poorer, less stable, less enjoyable than those of their parents and grandparents. I want them to know that there is hope—that their new-found voice is making a difference—and, whether through activism, community projects or building new, green businesses, they will change the world. But members of older generations still have a disproportionate influence. Younger people need to be given the chance to create a better world for themselves.

Above all, it's now clear that the tools required to avoid a climate disaster already exist, even if some of them require more research and development. Between deep, rapid emissions cuts and third-way technologies, we can do it. It really is over to you. Every one of you. We can now see the finishing line. Whether we win the race for a better future is your choice.

Afterword

A FEW weeks ago I dived the Great Barrier Reef, near Port Douglas. It was one of the saddest days of my life. I am haunted by what I've seen. And infuriated.

I had come with hope, for some recovery at least from the largest coral bleaching event on record. But what I found was worse than I could have imagined. The Great Barrier Reef is losing its adjective.

Most of the reef's usually vibrant staghorn and plate corals are covered with an ugly green slime. Even some of the massive stony corals—the hardiest of all—are scarred with the tell-tale white of bleaching. The reef's diverse and stunning fish population are starving.

A green turtle passes by. As the dead reef breaks down, its habitat will be eroded to rubble. And climate change is affecting the species in other ways. Rising seas have massively degraded its most important nesting site—Raine Island in the northern Great Barrier Reef. Those same rising waters caused, around 2011, the first mammal extinction brought about directly by climate change, when the entire habitat of the Bramble Key melomys (a native rodent unique to the Great Barrier Reef) was destroyed by saltwater intrusion.

As I reflected on my dive, I realised that I had been looking into the future. Because of el Nino, this year global temperatures rose by a third of a degree to 1.2C above the pre-industrial average. By the 2030s, this year's conditions will be average.

This great organism, the size of Germany and arguably the most diverse place on earth, is dying before our eyes. Having watched my father dying two years ago, I know what the signs of slipping away are. This is death, which ever-rising temperatures will allow no recovery from.

Unless we act now.

In December 2015, representatives from 195 nations arrived in Paris for the United Nations Climate Change Conference, technically known as COP 21—the twenty-first annual session of the 'Conference of the Parties' at an international climate convention. By the end of the conference, those nations had come together to sign the first universal, legally blinding global climate deal. It was a historic moment. And one of the biggest surprises of the climate meeting was the decision to include an ambition of capping the global temperature rise at 1.5°C, rather than the originally anticipated target of 2°C.

From a climate perspective, a world in which temperature rises are limited to 1.5°C is much preferable. It's possible that some of the Arctic ice cap may survive, and also, if the rise is gradual enough that some coral, maybe even on Australia's Great Barrier Reef, could adapt to the permanently warmer conditions, and that sea-level rise would be limited to a metre, if temperatures can be limited to 1.5°C.

In the wake of the Paris meeting, however, countries will be left pondering the actions required to do their part in limiting temperatures to 1.5°C. In the case of Australia and America, the actions required are often inconsistent with current policy and rhetoric.

The drastic nature of the actions required from all countries becomes evident from the fact that there is already enough greenhouse gas in the air to take temperatures to around 1.5°C by mid-century.

In carbon budget terms, we are already out of budget to reach a 1.5°C target.

Because we cannot immediately end the use of fossil fuels, if Australia and other nations are serious about reaching a 1.5°C target, they will need to cut emissions with unprecedented speed, as well as investing in third-way technologies to draw down whatever we put into the atmosphere in the future.

Because we don't yet have any technologies operating at a scale sufficiently large to draw down the amounts of CO_2 required, it makes sense to avoid putting as much greenhouse gas into the atmosphere as we can in the first place.

Whatever the pathway each nation chooses to honor its intention to limit temperature rise to 1.5°C, it will in most cases involve a dramatic shift in policy. And that shift cannot be delayed, because reaching a 1.5°C target is barely within our grasp in 2016.

A few years of inaction would make it entirely unreachable.

—Tim Flannery, 2016

Adapted from "The Great Barrier Reef is losing its adjective and it's our fault" in the Sydney Morning Herald *(Australia) and from "Tim Flannery: the biggest surprise from the Paris Climate Meeting" in the* New Daily *(Australia)*

Organisations Fighting for a Better Climate

Climate Communicators & Research Organisations

Carbon Tracker Initiative: carbontracker.org

Climate CoLab: climatecolab.org

The Climate Council: climatecouncil.org.au

The Climate Institute: climateinstitute.org.au

Climate Outreach and Information Network (COIN): climateoutreach.org.uk

ClimateWorks: climateworks.org

Doctors for the Environment Australia: dea.org.au

International Institute for Environment and Development: iied.org

Responding to Climate Change (RTCC): rtcc.org

Climate Advocacy

1 Million Women: 1millionwomen.com.au

Australian Conservation Foundation: acfonline.org.au

Avaaz: avaaz.org/en/index.php

CARE Climate Change: careclimatechange.org

Ceres: ceres.org

Citizens' Climate Lobby: citizensclimatelobby.org

The Climate Reality Project: climaterealityproject.org

Conservation International: conservation.org

Friends of the Earth: foe.org

Greenpeace: greenpeace.org/international/en

The Wilderness Society: wilderness.org.au

WWF: worldwildlife.org

Climate Action Networks

Climate Action Network: climatenetwork.org

Global Call for Climate Action: tcktcktck.org

Renewable Energy

Clean Energy Council: cleanenergycouncil.org.au

International Renewable Energy Agency (IRENA): irena.org

REN21—Renewably Energy Policy Network for the 21st Century: ren21.net

Solar Aid: solar-aid.org/about

Solar Citizens: solarcitizens.org.au

Divestment

350.org

Divest-Invest Philanthropy: divestinvest.org/philanthropy

Climate Youth Movements

Australian Youth Climate Coalition: aycc.org.au

Canadian Youth Climate Coalition: ourclimate.ca

China Youth Climate Action Network: cycan.org

UK Youth Climate Coalition: ukycc.org

Compiled by Max Newman, Australian Climate Council

Acknowledgements

Many people have contributed to this book through their discussions and encouragement. David Kent of HarperCollins Canada and Michael Heyward of Text Publishing, Australia, played an important role in convincing me that the book should be written. David Addison and Adepeju Adeosun of the Virgin Earth Challenge provided much encouragement, insight and editorial assistance. Alan Knight, while Director at Business in the Community, provided valuable input, including the suggestion that certain technologies should be called 'the third way'. Professors David Karoli and John Wiseman of Melbourne University, and Professor Will Steffen, Dr Martin Rice and Amanda McKenzie of the Australian Climate Council all read early drafts and provided invaluable commentary.

Endnotes

INTRODUCTION

1 '2014 Global Carbon Budget Released', Global Carbon Project, futureearth.org/news/2014-global-carbon-budget-released
Oskin, B., 'Global Carbon Emissions Reach New Record High', Live Science, 21 September 2014, livescience.com/47929-global-carbon-emissions-2014-record.html

2 'Global Energy-related Emissions of Carbon Dioxide Stalled in 2014', International Energy Agency, 13 March 2015, iea.org/newsroomandevents/news/2015/march/global-energy-related-emissions-of-carbon-dioxide-stalled-in-2014.html

3 The most recent year such figures are available.

4 Montzka, S. A. *et al.*, 'Non-CO_2 Gases and Climate Change', *Nature* 467, 43–50, 2011, nature.com/nature/journal/v476/n7358/abs/nature10322.html

5 *World Energy Outlook 2012*, International Energy Agency, iea.org/publications/freepublications/publication/english.pdf

6 'Global Energy-related Emissions of Carbon Dioxide Stalled in 2014', International Energy Agency, 13 March 2015, iea.org/newsroomandevents/news/2015/march/global-energy-related-emissions-of-carbon-dioxide-stalled-in-2014.html

7 'Fact Check: Do Australia, US Compare Favourably on Emissions Targets?', ABC News, 18 December 2014. abc.net.au/news/2014-12-18/greg-hunt-cherrypicking-emissions-reduction-targets/5896148

8 Parkinson, G., 'Abbott Blows his Carbon Budget in First Direct Action Auction', REnewEconomy, 23 April 2015, reneweconomy.com.au/2015/abbott-blows-his-carbon-budget-in-first-direct-action-auction-26282

CHAPTER ONE

1 Schmidt, G., 'The Emergent Patterns of Climate Change', May 2014, ted.com/talks/gavin_schmidt_the_emergent_patterns_of_climate_change/transcript

2 Neukom, R. *et al.*, 'Inter-hemispheric Temperature Variability over the Last Millennium', *Nature Climate Change*, 30 March 2014, nature.com/nclimate/journal/v4/n5/full/nclimate2174.html

3 'Climate Change not a Factor in NSW Bushfires: Tony Abbott', *Australian*, 23 October 2013, theaustralian.com.au/in-depth/bushfires/climate-change-not-a-factor-in-nsw-bushfires-tony-abbott/story-fngw0i02-1226745125303

4 Trenberth, K. E., 'Framing the Way to Relate Climate Extremes to Climate Change', *Climate Change* 115, 2, 283–90, 2012, link.springer.com/article/10.1007%2Fs10584-012-0441-5
Steffen, W., 'Quantifying the Strong Influence of Climate Change on Extreme Heat in

Australia', Australian Climate Council, 2015, climatecouncil.org.au/uploads/00ca18a19f f194252940f7e3c58da254.pdf

5 Chammas, M. and Donelly, B., 'Australian Open Tournament Referee Implements Extreme Heat Policy', *Sydney Morning Herald*, 16 January 2014, smh.com.au/sport/tennis/australian-open-tournament-referee-implements-extreme-heat-policy-20140116-30wl6.html

6 Knutson, T. *et al.*, 'Multimodal Assessment of Extreme Annual-Mean Warm Anomalies During 2013 over Regions of Australia and the Western Tropical Pacific', in Herring, M. P. *et al.*, 'Explaining Extreme Events of 2013 from a Climate Perspective', *Bulletin of the American Meteorological Society* (*BAMS*) 95, 9, September 2014, pp. 26–30.
Lewis, S. C. and Karoli, D. J., 'The Role of Anthropogenic Forcing in the Record 2013 Australia-wide Annual and Spring Temperatures', in Herring, M. P. *et al.*, *BAMS*, September 2014, pp. 31–34.
King, A. D. *et al.*, 'Climate Change Turns Australia's 2013 Big Dry into a Year of Record-Breaking Heat', in Herring, M. P. *et al.*, *BAMS*, September 2014, pp. 41–45.

7 'The New Climate Dice: Public Perception of Climate Change', NASA/GISS, 2012, giss.nasa.gov/research/briefs/hansen_17/

8 'Special Climate Statement 47—An Intense Heatwave in Central Eastern Australia', Bureau of Meteorology, 2014, bom.gov.au/climate/current/statements/scs47.pdf

9 Cooper, M., 'Death Toll Soared during Victoria's Heatwaves', *Age*, 6 April 2009.

10 Stott, P. A. *et al.*, 'Human Contribution to the European Heatwave of 2003', *Nature* 432, 610–14, 2 December 2004, nature.com/nature/journal/v432/n7017/abs/nature03089.html

11 Robine, J. *et al.*, 'Death Toll Exceeded 70,000 in Europe during the Summer of 2003', *Comptes Rendus Biologies*, 331, 2, 2008, pp. 171–78.

12 Stott, P. A. *et al.*, 'Human Contribution to the European Heatwave of 2003', *Nature* 432, 610–614, 2004.

13 'State of the Climate: Global Analysis, June 2010', National Climate Data Center, National Oceanic and Atmospheric Administration (NOAA), ncdc.noaa.gov/sotc/global/2010/6

14 climatecouncil.org.au/heatwaves-infographic

15 Melillo, J. M., Richmond, T., Yohe, G. (eds.), *Highlights of Climate Change Impacts in the United States: The Third National Climate Assessment*, US Global Change Research Program, 2014.

16 Barriopedro, D. *et al.*, 'The Hot Summer of 2010: Redrawing the Temperature Record Map of Europe', *Science* 332, 2011, pp. 220–24.

17 Price, C. and Rind, D., 'Possible Implications of Global Climate Change on Global Lightning Distributions and Frequencies', *Journal of Geophysical Research*, 99, D5, 10823–31, 1994.

18 Hughes, L., *Be Prepared: Climate Change and the NSW Bushfire Threat*, Australian Climate Council, 2014, climatecouncil.org.au/uploads/e8f0829c13a23f8a6b234962baadf419.pdf

19 Hughes, L. *et al.*, *Heatwaves: Hotter, Longer, More Often*, Australian Climate Council, 2014.

20 Kelly, R. *et al.*, 'Recent Burning of Boreal Forests Exceeds Fire Regime Limits of the Past 10,000 Years', *PNAS* 110, 32, 2013, pnas.org/content/110/32/13055.abstract

21 Kahn, B., 'Fires in NW Territories in Line with "Unprecedented" Burn', Climate Central, 17 July 2014, climatecentral.org/news/nw-fires-weather-climate-change-boreal-forests-17778

22 Hughes, L. *et al.*, *Heatwaves: Hotter, Longer, More Often*, Australian Climate Council, 2014.

23 Hughes, L. and Steffen, W., *Be Prepared: Climate Change and the Australian Bushfire Threat*, Australian Climate Council, 2014.

24 *Ibid.*

25 Westerling, A., *et al.*, 'Continued Warming Could Transform Greater Yellowstone Fire Regimes by Mid-21st Century', *PNAS* 108, 13165–170, 2011, pnas.org/content/108/32/13165.full.pdf+html

26 Westerling, A. *et al.*, 'Climate Change and Growth Scenarios for California Wildfire', *Climate Change* 109, 445–463, 2011, http://ulmo.ucmerced.edu/pdffiles/11cc_westerlingetal.pdf

27 Balshi, M. S. *et al.*, 'Assessing the Response of the Area Burned to Changing Climate in Western Boreal North America Using Multivariate Adaptive Regression Splines (MARS) Approach', *Global Change Biology* 15, 578–600, 2009.

28 Johnson, F. H., *et al.*, 'Estimated Global Mortality Attributable to Smoke from Landscape Fires', *Environmental Health Perspectives* 120, 695–701, 2012.

29 Spracklen, D. V. *et al.*, 'Impacts of Climate Change from 2000 to 2050 in Wildfire Activity and Carbonaceous Aerosol Concentrations in the Western United States', *Journal of Geophysical Research* 114, D20301, 2009.

30 Melillo, J. M., Richmond, T., Yohe, G. (eds.), *Highlights of Climate Change Impacts in the United States: The Third National Climate Assessment*, US Global Change Research Program, 2014.

31 Chen, Y., *et al.*, 'Evidence on the Impact of Sustained Exposure to Air Pollution on Life Expectancy from China's Huai River Policy', *PNAS*, 2013, pnas.org/content/110/32/12936.full

32 McMichael, A. J., 'Health Impacts in Australia in a Four Degree World', in Christoff, P. (ed.), *Four Degrees of Global Warming: Australia in a Hot World,* Routledge, Abingdon, 2014.

33 Melillo, J. M., *et al.* (eds.), *Highlights of Climate Change Impacts in the United States: The Third National Climate Assessment*, US Global Change Research Program, 2014.

34 *Ibid.*

35 *Ibid.*

36 Hughes, L. and McMichael, T., *The Critical Decade: Climate Change and Health*, Australian Climate Commission, Canberra, 2011.

37 *Ibid.*

38 Myers S. *et al.*, 'Increasing CO_2 Threatens Human Nutrition', *Nature* 510, 7503, 2014.

39 *China Daily*, 6–12 June 2014.

40 Haines, A. *et al.*, 'Climate Change and Human Health: Impacts, Vulnerability and Public Health', *London School of Hygiene and Tropical Medicine* 120, 7, 585–96, 2006. *See also* Hughes, L. and McMichael, T., *The Critical Decade: Climate Change and Human Health*, Australian Climate Commission, 2011, climatecouncil.org.au/commission-climate-change-and-health

41 McMichael, A. J., 'Climate Change: Health Risks Mount while Nero Fiddles', *Medical Journal of Australia*, 200, 9, 507–08, 2014, mja.com.au/journal/2014/200/9/climate-change-health-risks-mount-while-nero-fiddles

42 McMichael, A. J., 'Health Impacts in Australia in a Four Degree World', in: Christoff, P. (ed.), *Four Degrees of Global Warming: Australia in a Hot World,* Routledge, Abingdon, 2014.

43 *Ibid.*

CHAPTER TWO

1 O'Malley, M., 'Warming Seas Wash Away Some of South Florida's Glitz', *Sydney Morning Herald*, 21 December 2014.

2 Melillo, J. M., Richmond, T., Yohe, G. (eds.), *Highlights of Climate Change Impacts in the United States: The Third National Climate Assessment*, US Global Change Research Program, 2014.

3 'Special Climate Statement 44: Extreme Rainfall and Flooding in Coastal Queensland and New South Wales', Bureau of Meteorology, 2013, bom.gov.au/climate/current/statements/scs44.pdf

4 'State Disaster Group Annual Report 2011–2012', Queensland Government, disaster.qld.gov.au/Disaster-Resources/Documents/State%20Disaster%20Management%20Group%20Annual%20Report%202011-2012.pdf

5 'Record Sea Surface Temperatures', Bureau of Meteorology, bom.gov.au/climate/enso/history/ln-2010-12/SST-records.shtml

6 Christidis, N. *et al.*, 'An Attribution Study of the Heavy Rainfall over Eastern Australia in March 2012', *BAMS* 19, 58–61, 2013.
King, A. D. *et al.*, 'Limited Evidence of Anthropogenic Influence on the 2011–12 Rainfall over Southeast Australia', *BAMS* 19, 55–58, 2013.
Lewis, S. C. and Karoli, D. J., 'Are Estimates of Anthropogenic and Natural Influences on Australia's Extreme 2010–2012 Rainfall Model Dependent?' *Climate Dynamics*, 1–17, 2014.

7 Wenju Cai *et al.*, 'Increased Frequency of Extreme La Niña Events under Greenhouse Warming', *Nature Climate Change* 5, 132–37, 2015, nature.com/nclimate/journal/v5/n2/full/nclimate2492.html?WT.ec_id=NCLIMATE-201502

8 Howe, C. and Rogers, E., 'Queensland's Largest Drought-declared Area Ever', ABC News, 11 March, 2014, abc.net.au/news/2014-03-07/qld-drought--most-widespread-ever-recorded/5306044

9 Melillo, J. M., Richmond, T., Yohe, G. (eds.), *Highlights of Climate Change Impacts in the United States: The Third National Climate Assessment*, US Global Change Research Program, 2014.

10 'Rainfall, Dam Storage and Water Supply', Water Corporation, Perth, www.watercorporation.com.au/water-supply-and-services/rainfall-and-dams/streamflow/streamflowhistorical

11 'Desalination', Water Corporation, Perth, www.watercorporation.com.au/water-supply-and-services/solutions-to-perths-water-supply/desalination

12 Thompson, A., 'Bleak California Snowpack "Obliterates" Record Low', Climate Central, 1 April 2015, climatecentral.org/news/california-snowpack-obliterates-record-low-18847

13 Magill, B., 'Epic Drought in West Is Literally Moving Mountains', Climate Central, 21 August 2014, climatecentral.org/news/epic-drought-in-west-is-moving-mountains-17924

14 'DWR Increases 2015 Water Allocation to Water Contractors', California Department of Water Resources., press release, 15 January 2015, water.ca.gov/news/newsreleases/2015/011515increases.pdf

15 Burt, C. C., 'California Drought Update—May 2014', Weather Underground, 2014, wunderground.com/blog/weatherhistorian/comment.html?entrynum=270

16 Cook, B. I. *et al.*, 'Unprecedented 21st Century Drought Risk in the American Southwest and Central Plains', *Science Advances* 1, 1, 12 February 2015, advances.sciencemag.org/content/1/1/e1400082

17 Amos, J., 'US "at Risk of Mega-drought Future"', BBC News, 13 February 2015, bbc.com/news/science-environment-31434030

18 Rotstayn, L. D. and Lohmann, U., 'Tropical Rainfall Trends and the Indirect Rainfall Effect', *Journal of Climate* 15, 2103–16, 2002.

19 Held, I. M., Delworth, T. L. *et al.*, 'Simulation of Sahel Drought in the 20th and 21st Centuries', *PNAS* 102 (50), 17891–96, 2005

20 Zhang, R., Delworth, T. L., 'Impact of Atlantic Multidecadal Oscillations on India/Sahel Rainfall and Atlantic Hurricanes', *Geophysical Research Letters*, 33 (17), 2006.

21 IPCC Working Group, *Climate Change 2013: The Physical Science Basis*, 1. 14. 6., ipcc.ch/report/ar5/wg1/

22 Melillo, J. M., Richmond, T., Yohe, G. (eds.), *Highlights of Climate Change Impacts in the United States: The Third National Climate Assessment*, US Global Change Research Program, Washington DC, 2014.

23 'Arctic Sea Ice and News Analysis', National Snow & Ice Data Center, 7 April 2015, nsidc.org/arcticseaicenews/

24 *Climate Change 2001*, IPCC, ipcc.ch/ipccreports/tar

25 'Arctic Sea Ice Reaches Lowest Maximum Extent on Record', National Snow & Ice Data Center, 19 March 2015, nsidc.org/arcticseaicenews/

26 Kohout, A. L. *et al.*, 'Storm-induced Sea-ice Breakup and the Implications for Ice Extent', *Nature* 509, 604–607, 29 May 2014, nature.com/nature/journal/v509/n7502/

full/nature13262.html

27 Williams, G., 'Why Is Arctic Sea Ice Growing?', *The Conversation*, 29 October 2013, theconversation.com/why-is-antarctic-sea-ice-growing-19605

28 McMillan, M. *et al.*, 'Increased Ice Losses from Antarctica Detected by CryoSat-2', *Geophysical Research Letters* 41, 3899–905, 2014, onlinelibrary.wiley.com/doi/10.1002/2014GL060111/abstract

29 Favier, L. *et al.*, 'Retreat of Pine Island Glacier Controlled by Marine Ice Sheet Instability', *Nature Climate Change* 4, 117–21, 2014, nature.com/nclimate/journal/vaop/ncurrent/full/nclimate2094.html

30 Lynch, P., 'The "Unstable" West Antarctic Ice Sheet: A Primer', NASA, 12 May 2014, jpl.nasa.gov/news/news.php?release=2014-147

31 Favier, L. *et al.*, 'Retreat of Pine Island Glacier Controlled by Marine Ice-sheet Instability', *Nature Climate Change* 4, 117–21, 2014.

32 Lynch, P., 'The "Unstable" West Antarctic Ice Sheet: A Primer', NASA, 12 May 2014.

33 Spence, P. *et al.*, 'Rapid Subsurface Warming and Circulation Changes of Antarctic Coastal Waters by Poleward Shifting Wind', *Geophysical Research Newsletters* 41, 13 4601–10, 16 July 2014.

CHAPTER THREE

1 Glickson, A. Y., *Evolution of the Atmosphere, Fire and the Anthropocene Climate Event Horizon*, Springer, Dordrecht, 2014.

2 *Ibid.*

3 Kolbert, E., *The Sixth Extinction: An Unnatural History*, Henry Holt and Company, New York, 2014.

4 Langdon, C. *et al.*, 'Seawater Carbonate Chemistry and Processes during an Experiment with Coral Reef', 2000, doi.pangaea.de/10.1594/PANGAEA.721195

5 McCalman, I., *The Great Barrier Reef: A Passionate History*, Penguin Books, Melbourne, 2013.

6 Kolbert, E., 'The Acid Sea', *National Geographic*, 11 April 2011, ngm.nationalgeographic.com/2011/04/ocean-acidification/kolbert-text

7 Hoegh-Guldberg, O. *et al.*, 'Coral Reefs under Rapid Climate Change and Ocean Acidification', *Science* 318, 1737–42, 2007.

8 Edmunds, P. J. *et al.*, 'Understanding the Threats of Ocean Acidification to Coral Reefs', *Oceanography* 26 (3), 149–52, 2013, doi.org/10.5670/oceanog.2013.57

9 Campbell, A. L. *et al.*, 'Ocean Acidification Increases Copper Toxicity to the Early Life History Stages of the Polychaete *Arenicola marina* in Artificial Seawater', *Environmental Science and Technology* 48, 16, 9745–53, 2014, pubs.acs.org/doi/abs/10.1021/es502739m

10 Hoffmann, L. J. *et al.*, 'Influence of Ocean Warming and Acidification on Trace Metal Biogeochemistry', *Marine Ecology Progress Series* 470, 191–205, 2012, int-res.com/articles/theme/m470p191.pdf

11 Harrabin, R., 'Science Chief Warns on Acid Oceans', BBC News, 24 October 2014,

bbc.co.uk/news/science-environment-29746880

12 Jiang, Z. *et al.*, 'Influence of Seaweed Aquaculture on Marine Inorganic Carbon Dynamics and CO_2 Sea-Air Flux', *Journal of the World Aquaculture Society* 44, 133–40, 2013.

13 Han, T. *et al.*, 'Carbon Dioxide Fixation by the Seaweed *Gracilaria lemaneiformis* in Integrated Multi-trophic Aquaculture with the Scallop *Chlamys farreri* in Sanggou Bay, China', *Aquaculture International* 21, 1035–43, 2013.

14 Lenstra, J. *et al.*, 'Economic Aspects of Open Ocean Seaweed Cultivation', Energy Research Centre of the Netherlands, ecn.nl/docs/library/report/2011/m11100.pdf

15 N'Yeurt, A. *et al.*, 'Negative Carbon via Ocean Afforestation', *Process Safety and Environmental Protection* 90, 467–74, 2012.

CHAPTER FOUR

1 Leakey, R. and Lewin, R., *The Sixth Extinction: Patterns of Life and the Future of Humankind*, Doubleday, New York, 1995.

2 McCalman, I., *The Great Barrier Reef: A Passionate History*, Penguin Books, Melbourne, 2013.

3 De'ath, G. *et al.*, 'The 27-Year Decline of Coral Cover on the Great Barrier Reef and Its Causes', *PNAS* 109, 17995–99, 2012.

4 Hoegh-Guldberg, O. *et al.*, 'Australia's Marine Resources in a Warm, Acid Ocean', in Christoff, P. (ed.), *Four Degrees of Global Warming: Australia in a Hot World*, Routledge, Abingdon, 2013, p 92.

5 *Ibid.*

6 Stempniewicz, L. *et al.*, 'Unusual Feeding and Hunting Behaviours of Polar Bears on Spitsbergen', *Polar Record* 50, 2, 216–19, 2013.

7 Obbard, M. E. *et al.* (eds.), *Polar Bears: Proceedings of the 15th Working Meeting of the IUCN/SSC Polar Bear Specialist Group, Copenhagen, Denmark, 29 June–3 July 2009*, Gland, Switzerland and Cambridge, IUCN, 2010.

8 polarbearscience.com/2013/08/07/ian-stirlings-latest-howler-the-polar-bear-who-died-of-climate-change/

9 Mooallem, J., *Wild Ones*, Penguin, New York, 2013.

10 Amstrup, S. C. *et al.*, *USGS Science Strategy to Support U.S. Fish and Wildlife Service Polar Bear Listing Decision: Forecasting the Range-wide Status of Polar Bears at Selected Times in the 21st Century*, US Geological Survey, Reston, Virginia, 2007.

11 *Ibid.*

12 Ducklow, H. W. *et al.*, 'West Antarctic Peninsula: An Ice-dependent Coastal Marine Ecosystem in Transition', *Oceanography* 26, 3, 190–202, 2013.

13 *Ibid.*

14 Fraser, W. R. *et al.*, 'A Non-marine Source of Variability in Adelie Penguin Demography', *Oceanography* 26, 3, 207–09, 2013.

15 Williams, S. E. *et al.*, 'Climate Change in Australian Tropical Rainforests: An Impending Environmental Catastrophe', *Proceedings of the Royal Society* (B) 270, 1887–92, 2003.

16 National Wildlife Federation, 'Global Warming and the American Pika', nwf.org/Wildlife/Threats-to-Wildlife/Global-Warming/Effects-on-Wildlife-and-Habitat/Pika.aspx

17 Woody, T. 'The American Pika Could Survive Climate Change by Eating Its Own Feces', *Quartz*, 2013, qz.com/160506

18 Melillo, J. M., Richmond, T., Yohe, G. (eds.), *Highlights of Climate Change Impacts in the United States: The Third National Climate Assessment*, US Global Change Research Program, 2014.

19 'The IUCN Red List of Threatened Species', iucnredlist.org

20 IPCC Working Group II, 'Climate Change 2014: Impacts, Adaptation, and Vulnerability', ipcc-wg2.gov/AR5/

21 Kolbert, E., *The Sixth Extinction: An Unnatural History*, Henry Holt and Co, New York, 2014.

22 Pimm, S. L. *et al.*, 'The Biodiversity of Species and their Rates of Extinction, Distribution, and Protection', *Science* 344, 6187, 2014.

CHAPTER FIVE

1 Stocker, T. F. *et al.* (eds.), *Working Group I Contribution to the IPCC Fifth Assessment Report Climate Change 2013: The Physical Science Basis, Summary for Policymakers*, IPCC, 2013, ipcc.ch/pdf/assessment-report/ar5/wg1/WGIAR5_SPM_brochure_en.pdf

2 Glickson, A. Y., *Evolution of the Atmosphere, Fire and the Anthropocene Climate Event Horizon*, Springer, Dordrecht, 2014.

3 *Ibid.*

4 Christoff, P. (ed)., *Four Degrees of Global Warming: Australia in a Hot World*, Routledge, Abingdon, 2014.

5 Schiermeier, Q., 'Atlantic Current Strength Declines', *Nature*, 13 May 2014, nature.com/news/atlantic-current-strength-declines-1.15209

6 'Understanding Climate Change Impacts on the Amazon Rainforest', metoffice.gov.uk/research/news/amazon-dieback

7 Bagley, J. E. *et al.*, 'Drought and Deforestation: Has Land Cover Change Influenced Recent Precipitation Extremes in the Amazon?', *American Meteorological Society* 27, 345–61, 2014.

8 cemaden.gov.br

9 Davies, W., 'Brazil Drought: São Paulo Sleepwalking into Water Crisis', BBC News, 7 November 2014, bbc.co.uk/news/world-latin-america-29947965

10 'Methane Budget Highlights', Global Carbon Project, globalcarbonproject.org/methanebudget/13/hl-compact.htm

11 Kerr, R. A., 'Gas Hydrate Resource: Smaller but Sooner', *Science* 303, 946–47, 2004.

12 Benton, M. J. *et al.*, *When Life Nearly Died: The Greatest Mass Extinction of All Time*, Thames & Hudson, London, 2003.

13 *IPCC Fifth Assessment Report*, Chapter 6, ipcc.ch/report/ar5/

14 Marshall, M., 'As Arctic Ocean Warms, Megatonnes of Methane Bubble Up', *New Scientist*, 17 August, 2009, newscientist.com/article/dn17625-as-arctic-ocean-warms-megatonnes-of-methane-bubble-up.html#.VEz5VRakzxw

15 'Massive Methane Concentrations over the Laptev Sea', *Arctic News*, 22 February 2014, arctic-news.blogspot.com.au/2014/02/massive-methane-concentrations-over-the-laptev-sea.html

16 Moskvitch, K., 'Mysterious Siberian Crater Atributed to Methane', *Nature*, 31 July 2014, nature.com/news/mysterious-siberian-crater-attributed-to-methane-1.15649

17 'Methane Bubbles Climate Trouble', BBC News, 7 September 2006, news.bbc.co.uk/2/hi/science/nature/5321046.stm

18 Jerew, B., 'Arctic Methane Could Accelerate Climate Change', The Green Optimistic, greenoptimistic.com/artic-methane-accelerate-climate-change-20140811/#.VWaZW6acvyh

19 Kirschke, S. *et al.*, 'Three Decades of Global Methane Sources and Sinks', *Nature Geoscience*, 22 September 2013.

20 Global Carbon Project, 'Methane Budget Highlights', globalcarbonproject.org/methanebudget/13/hl-compact.htm

CHAPTER SIX

1 *QandA*, ABC TV, 24 June 2014, abc.net.au/tv/qanda/txt/s4008999.htm

2 'Australian GDP Growth and Inflation', Reserve Bank of Australia, 4 March 2015, rba.gov.au/chart-pack/au-gdp-growth.html

3 'Germany GDP Growth Rate, 1991–2015', Trading Economics, tradingeconomics.com/germany/gdp-growth

4 Glenday, J. and Griffiths, E., 'Tony Abbott Arrives in US Where Climate Stance Will Be under Scrutiny', ABC News, 10 June 2014, abc.net.au/news/2014-06-10/tony-abbott-finds-friend-in-canadian-pm-harper-over-carbon-tax/5511252

5 Freedman, A., 'Science Historian Reacts to Hacked Climate E-mails', *Washington Post*, 23 November 2009.

6 Souder, W. B., *On a Farther Shore*: *The Life and Legacy of Rachel Carson*, Crown Publishing, New York, 2012.

7 Raworth, K., 'Hunting for Green Growth in the G20', Oxfam, 20 January 2012, policy-practice.oxfam.org.uk/blog/2012/01/hunting-for-green-growth-in-the-g20

8 'Electricity Emissions Update—Data to 31 March 2015', CEDEX, Carbon Emissions Index, www.pittsh.com.au/assets/files/Cedex/CEDEX%20Electricity%20Update%20April%202015.pdf

9 Ting, I., 'Green Economy Index 2014: Australia Ranked Last for Leadership', *Sydney Morning Herald*, 21 October 2014, smh.com.au/business/carbon-economy/

green-economy-index-2014-australia-ranked-last-for-leadership-20141020-118slt.html#ixzz3Gmql0k7Z

CHAPTER SEVEN

1 'Coal's Share of Global Energy Mix to Continue Rising, with Coal Closing in on Oil as World's Top Energy Source by 2017', International Energy Agency, 17 December 2012, iea.org/newsroomandevents/pressreleases/2012/december/name,34441,en.html
2 'World Energy Outlook 2012', Executive Summary, International Energy Agency, iea.org/publications/freepublications/publication/English.pdf
3 Buckley, T., 'India's Plan to Stop Importing Coal Deals another Blow to Australia', RenewEconomy, 13 November 2014, reneweconomy.com.au/2014/indias-plan-stop-importing-coal-deals-another-blow-australia-68894
4 'Turkish Coalmine Disaster: Final Death Toll from Soma Accident Stands at 301', ABC News, 18 May 2014, abc.net.au/news/2014-05-17/turkey-coalmine-collapse-fire-delays-rescue-work/5459882
5 Guay, J., 'China's Coal Consumption Has Finally Decreased', CleanTechnica, 26 August 2014, cleantechnica.com/2014/08/26/chinas-coal-consumption-finally-decreased
6 *Ibid.*
7 McGarrity, J., 'China's Coal Output Falls for First Time this Century', China Dialogue, 27 January 2015, europeanclimate.us9.list-manage.com/track/click?u=49de5e289696467538e9c8e65&id=38d335e3c0&e=bdeb1feab7
8 Bhattacharya, A., 'China's Coal Tariff Prolongs the Pain', *Wall Street Journal*, 10 October 2014, wsj.com/articles/chinas-coal-tariff-prolongs-the-pain-heard-on-the-street-1412937687
9 'China's Thermal Coal Demand Expected to Grow', World Coal, 4 June 2013, worldcoal.com/news/coal/articles/Chinese_thermal_coal_demand_will_reach_7_billion_tpa_214.aspx#.VDYt5Bakzxw
10 Pearse, G., *High & Dry: John Howard, Climate Change and the Selling of Australia's Future*, Penguin, Melbourne, 2007.
11 Heber, A., 'ICAC Says Coal Corruption Was Inevitable, Recommends Tighter Controls', Australian Mining, 30 October 2013, miningaustralia.com.au/news/icac-says-coal-corruption-was-inevitable-recommend
12 Gerathy, S., 'ICAC: Buildev Wanted to Discuss Newcastle Coal Loader Plans with Mike Baird, Inquiry Told', ABC News, 28 August 2014, abc.net.au/news/2014-08-28/icac-told-buildev-wanted-to-meet-premier-baird/5702922
13 'Indian Supreme Court Cancels 214 Coal Scandal Permits', BBC News, 24 September 2014, bbc.co.uk/news/world-asia-india-29339842
14 'Carbon Supply Cost Curves: Evaluating Financial Risk to Coal Capital Expenditures', Carbon Tracker Initiative, 22 September 2014, carbontracker.org/wp-content/uploads/2014/09/CTI-Coal-report-Sept-2014-WEB1.pdf
15 Paton, J. *et al.*, 'Coal Assets Abound for Those Willing to Bet on Rebound: Real

M&A', *Bloomberg Business*, 11 March 2015, bloomberg.com/news/articles/2015-03-10/coal-assets-abound-for-those-willing-to-bet-on-rebound-real-m-a

16 Locke, S., 'Ten Thousand Coal Mining Jobs Gone in 2 Years', ABC Rural, 19 May 2014, abc.net.au/news/2014-05-16/coal-outlook/5462226

17 'Carbon Supply Cost Curves: Evaluating Financial Risk to Coal Capital Expenditures', Carbon Tracker Initiative, 22 September 2014, carbontracker.org/wp-content/uploads/2014/09/CTI-Coal-report-Sept-2014-WEB1.pdf

18 Seccombe, M., 'Contrary in the Coalmine', *Saturday Paper*, 22–28 November 2014.

19 Seccombe, M., 'The End of Coal', *Saturday Paper*, 26 April 2014, thesaturdaypaper.com.au/news/resources/2014/04/26/the-end-coal/1398434400#.VTszKZPcho4

20 *Ibid.*

21 Parkinson, G., 'Coal-fired Generation in US to Fall by ¼ by 2020', RenewEconomy, 15 September 2015, reneweconomy.com.au/2014/coal-fired-generation-in-us-to-fall-by-14-by-2020-2020

22 Carr, M., 'EU Shutters Most Coal, Natural Gas Power in Six Years', *Bloomberg Business*, 11 February 2015, bloomberg.com/news/articles/2015-02-11/eu-shutters-most-coal-natural-gas-power-in-six-years

23 'Carbon Supply Cost Curves: Evaluating Financial Risk to Coal Capital Expenditures', Carbon Tracker Initiative, carbontracker.org/wp-content/uploads/2014/09/CTI-Coal-report-Sept-2014-WEB1.pdf

24 *Mining Coal, Mounting Costs: The Life Cycle Conspequences of Coal*, Center for Health and the Global Environment, Harvard Medical School, Boston, 2011, chge.med.harvard.edu/sites/default/files/resources/MiningCoalMountingCosts.pdf

25 Schneider, C. and Banks, J., 'The Toll from Coal', Clean Air Taskforce, September 2010, catf.us/resources/publications/files/The_Toll_from_Coal.pdf

26 *Mining Coal, Mounting Costs: The Life Cycle Conspequences of Coal*, Center for Health and the Global Environment, Harvard Medical School, Boston, 2011.

CHAPTER EIGHT

1 'Oil Falls to Near Record Low', *Sydney Morning Herald*, 13 January 2015, smh.com.au/business/oil-falls-to-nearrecord-low-20150112-12mw4p.html

2 'Unsustainable Energy', *Economist*, 11–17 October 2014.

3 Nelder, C., 'The Clean Energy Transition Is Unstoppable, so Why Fight It?', ZD Net, 18 April 2014, smartplanet.com/blog/the-take/clean-energy-transition-unstoppable-so-why-fight-it/

4 'Unsustainable Energy', *Economist*, 11–17 October 2014.

5 Arnsdofr, I., 'Falling Demand for Oil Is the Biggest Concern for Saudis', *Bloomberg Business*, 26 January 2015, bloomberg.com/news/articles/2015-01-26/for-saudis-falling-demand-for-oil-is-the-biggest-concern

6 'Renewable Jet Fuels—Carbon War Room', vimeo.com/39042259

7 *Ibid.*

8 Stephan, N., 'Fossil Fuel Subsidies Fall in Gain for Renewables', *Bloomberg Business*, 30 January 2015, bloomberg.com/news/articles/2015-01-30/fossil-fuel-subsidies-fall-in-gain-for-renewables

9 Nelder, C., 'The Clean Energy Transition Is Unstoppable, so Why Fight It?', ZD Net, 18 April 2014.

10 Johnston, R., 'The Problem with Condensates', *OilPrice*, 18 February, 2014, oilprice.com/Energy/Energy-General/The-Problem-With-Condensates.html

CHAPTER NINE

1 Kirkpatrick, R., *The Nummulosphere*, Lam Ley & Co., London, 1916 (Vols 1–2); William Clowes & Sons, London,1917, (Vol 3).

2 'McKinsey on Sustainability and Resource Productivity', McKinsey & Company, August 2014, (copies available on request from McKinsey_on_SRP@McKinsey.com).

3 Helm, D., *The Carbon Crunch: How We're Getting Climate Change Wrong—And How to Fix It*, Yale University Press, Princeton, 2012.

4 MacDonald-Smith, A., 'Brisbane Spot Gas Price Hits Record Low Near Zero', *Sydney Morning Herald*, 3 October 2014, smh.com.au/business/mining-and-resources/brisbane-spot-gas-price-hits-record-low-near-zero-20141002-10pckw.html

5 Mills, L., 'Global Trends in Clean Energy Investment', *Bloomberg New Energy Finance*, 9 January 2015, about.bnef.com/presentations/clean-energy-investment-q4-2014-fact-pack/content/uploads/sites/4/2015/01/Q4-investment-fact-pack.pdf

6 Parkinson, G., 'Citigroup: How Solar Module Prices Could Fall to 25c/Watt', RenewEconomy, 1 April 2013, reneweconomy.com.au/2013/citigroup-how-solar-module-prices-could-fall-to-25cwatt-41384

7 Pers. comm., May 2014.

8 Levant, E., *Groundswell: The Case for Fracking*, Random House, Toronto, 2014.

9 Powers, B., *Cold, Hungry and in the Dark: Exploding the Natural Gas Supply Myth*, New Society Publishers, New York, 2013.

10 Martin, C., 'Shale Gas Boom Leaves Wind Companies Seeking More Subsidy', *Bloomberg Business*, 8 April 2014, bloomberg.com/news/articles/2014-04-06/shale-gas-boom-leaves-wind-companies-seeking-more-subsidy

11 Trabish, H. K., 'Experts: The Cost Gap Between Renewables and Natural Gas "Is Closing"', *Greentech Media*, 6 May 2014, greentechmedia.com/articles/read/The-Price-Gap-Is-Closing-Between-Renewables-and-Natural-Gas

12 Martin, C., 'Shale Gas Boom Leaves Wind Companies Seeking More Subsidy', *Bloomberg Business*, 8 April 2014.

13 *Ibid.*

14 Nelder, C., 'The Clean Energy Transition Is Unstoppable, so Why Fight It?', ZD Net, 18 April 2014.

15 Jeremy Leggett, email to supporters, 22 April 2014.

16 McJeon, H. *et al.*, 'Limited Impact on Decadal Scale Climate Change from Increased Use of Natural Gas', *Nature* 514, 482–85, 23 October 2014, nature.com/nature/journal/v514/n7523/abs/nature13837.html

CHAPTER TEN

1 Steffen, W., 'Unburnable Carbon: Why We Need to Leave Fossil Fuels in the Ground', Climate Council, 23 April 2015, climatecouncil.org.au/unburnable-carbon-why-we-need-to-leave-fossil-fuels-in-the-ground

2 *Ibid.*

3 *Ibid.*

4 McKibben, B. and Naidoo, K., 'Do the Math: Fossil Fuel Investments Add up to Chaos', Fossil Free, 1 November 2013, gofossilfree.org/do-the-math-fossil-fuel-investments-add-up-to-climate-chaos/

5 'Unburnable Carbon: Are the World's Financial Markets Carrying a Carbon Bubble?', Carbon Tracker, 2014, carbontracker.org/wp-content/uploads/2014/09/Unburnable-Carbon-Full-rev2-1.pdf.

6 *Ibid.*

7 'Measuring the Global Fossil Fuel Divestment Movement', Arabella Advisors, 2014, arabellaadvisors.com/wp-content/uploads/2014/09/Measuring-the-Global-Divestment-Movement.pdf

8 *Ibid.*

9 Weissman, J., 'Norway's Giant Oil Fund Will Divest from Coal, Irony Noted', Moneybox, 28 May 2015, slate.com/blogs/moneybox/2015/05/28/norway_s_oil_fund_will_divest_from_coal_yes_this_is_kind_of_ironic.html

10 'Fund Rankings', Sovereign Wealth Fund Institute, swfinstitute.org/fund-rankings/

11 Flood, C., 'Fossil Fuel Divestment Action Gathers Momentum', *Financial Times*, 4 May 2015, ft.com/intl/cms/s/0/6ee5672e-edb0-11e4-987e-00144feab7de.html#axzz3ZL1pTeIP

12 McGlade, C. and Ekins, P., 'The Geographical Distribution of Fossil Fuels Unused when Limiting Global Warming to 2 Degrees C', *Nature* 517, 187–90, 2015, nature.com/nature/journal/v517/n7533/full/nature14016.html

13 Biello, D., 'Where in the World Are the Fossil Fuels that Cannot Be Burned to Restrain Global Warming?', *Scientific American*, 7 January 2015.

14 'Carbon Risk Beyond Regulation', S&P–Reputex, 23 May 2014, reputex.com/publications/latest-research/report-standard-poors-reputex-corporate-carbon-risks-extend-beyond-regulated-entities/

15 *Ibid.*

16 Beckman, K., 'Interview Fatih Birol, IEA: "Gradual change will not save us"', *Energy Post*, 3 June 2014, energypost.eu/interview-fatih-birol-iea-gradual-change-will-save-us/

17 'McKinsey on Sustainability and Resource Productivity', 2, 2014, mckinsey.com/client_service/sustainability/latest_thinking/mckinsey_on_sustainability

18 *Ibid.*

19 Kenny, T., 'What Are Green Bonds?', *About Money*, November 2014, bonds.about.com/od/munibonds/a/What-Are-Green-Bonds.htm

20 Lubber, M., 'Hope for Clean Economy as $20bn in Green Bonds Are Issued in 2014', *Guardian*, 12 September 2014, theguardian.com/sustainable-business/2014/sep/11/rise-green-bonds-hope-clean-economy-climate-change

21 McGrath, M., '"Carbon Bubble" Threatens Stock Markets, say MPs', BBC News, 6 March 2014, bbc.com/news/science-environment-26455763

22 *Ibid.*

CHAPTER ELEVEN

1 Gourmellon, G., 'Wind, Solar Generation Capacity Catching Up with Nuclear Power', Worldwatch Institute, 30 September 2014, worldwatch.org/node/144

2 *Ibid.*

3 *Ibid.*

4 'Nuclear Power in China', World Nuclear Association, 30 April 2015, world-nuclear.org/info/Country-Profiles/Countries-A-F/China--Nuclear-Power/

5 Haider, M., '32 Nuclear Plants to Produce 40,000MW: PAEC', *The News*, 27 February 2014, thenews.com.pk/Todays-News-3-235039-32-nuclear-plants-to-produce-40,000MW:-PAEC

6 'Nuclear Power in Russia', World Nuclear Association, April 2015, world-nuclear.org/info/Country-Profiles/Countries-O-S/Russia--Nuclear-Power/

7 'Nuclear Power in Canada', World Nuclear Association, February 2015, world-nuclear.org/info/Country-Profiles/Countries-A-F/Canada--Nuclear-Power/

8 Nagata K., 'Fukunshima Meltdowns Set Nuclear Energy Debate on its Ear', *Japan Times*, 3 January 2012.

9 'Why Is Nuclear Energy Necessary in Japan?', FEPC, www.fepc.or.jp/english/nuclear/power_generation/plants/index.html

10 'Fukushima's Impact on Japan's Economy Three Years On', BBC News, 11 March 2014, bbc.com/news/business-26524084

11 'Nuclear Power in Germany', World Nuclear Association, February 2015, world-nuclear.org/info/Country-Profiles/Countries-G-N/Germany

12 Carnegy, H., 'France to Set Nuclear Power Cap', *Financial Times*, 18 June 2014, ft.com/intl/cms/s/0/0ac6dc96-f6e4-11e3-8ed6-00144feabdc0.html#axzz3YIuK875C

13 'Nuclear Waste Disposal Concepts', World Nuclear Association, March 2015, world-nuclear.org/info/Nuclear-Fuel-Cycle/Nuclear-Wastes/International-Nuclear-Waste-Disposal-Concepts/

CHAPTER TWELVE

1 Lash, W. H., 'A Current View of the Kyoto Climate Change Treaty', Center for the

Study of American Business, Washington University, St Louis, 1999.

2 Kelly-Detweiler, P., 'Texas Sets New Wind Power Record', *Forbes Magazine*, 30 March 2014, forbes.com/sites/peterdetwiler/2014/03/30/texas-sets-new-wind-power-record/

3 Greene, N., 'Renewables Beat Fossil Fuels for Second Year in a Row', EcoWatch, 21 April 2015, ecowatch.com/2015/04/21/renewables-beat-fossil-fuels-bloomberg/

4 Parkinson, G., 'Fossil Fuels to Be Stranded by Economics, Innovation and Climate', RenewEconomy 21 April 2015, reneweconomy.com.au/2015/fossil-fuels-to-be-stranded-by-economics-innovation-and-climate-70623

5 'How to Lose Half a Trillion Euros', *Economist*, 12 October 2013, economist.com/news/briefing/21587782-europes-electricity-providers-face-existential-threat-how-lose-half-trillion-euros

6 '20% Wind Energy by 2030: Increasing Wind Energy's Contribution to U.S. Electiricity Supply', U.S. Department of Energy, July 2008, nrel.gov/docs/fy08osti/41869.pdf

7 'McKinsey on Sustainability & Resource Productivity', McKinsey & Company, July 2014, mckinsey.com/client_service/sustainability/latest_thinking/mckinsey_on_sustainability

8 'Deutsche Bank's 2015 Solar Outlook: Acceleration Investment and Cost Competitiveness', Deutsche Bank, January 2015, db.com/cr/en/concrete-deutsche-banks-2015-solar-outlook.htm

9 'A New Spin on Production', *Pictures of the Future*, Siemens, Spring 2013, siemens.com/innovation/apps/pof_microsite/_pof-spring-2013/_pdf/en/A_new_spin_on_production_EN.pdf

10 Trancik, J. E., 'Back the Renewables Boom', *Nature* 507, 300, 2014, nature.com/news/renewable-energy-back-the-renewables-boom-1.14873

11 *Ibid.*

12 Musk, E. *et al.*, 'Solar at Scale', SolarCity blog, 16 June 2014, blog.solarcity.com/silevo

CHAPTER THIRTEEN

1 Wesoff, E., 'Tesla CTO on Energy Storage: "We Should All be Thinking Bigger"', GreenTechMedia, 27 May 2014, greentechmedia.com/articles/read/Tesla-CTO-on-Energy-Storage-We-Should-All-Be-Thinking-Bigger

2 'Gigafactory', Tesla, teslamotors.com/sites/default/files/blog_attachments/gigafactory.pdf

3 Nelder, C., 'The Clean Energy Transition Is Unstoppable, so Why Fight It?', ZDNet, 18 April 2014.

4 Wesoff, E., 'Tesla CTO on Energy Storage: "We Should All be Thinking Bigger"', GreenTechMedia, 27 May 2014.

5 Jenkins, J., 'Cost of Batteries for Electric Vehicles Falling More Rapidly than Projected', The Energy Collective, 13 April 2015, theenergycollective.com/jessejenkins/2215181/cost-batteries-electric-vehicles-falling-more-rapidly-projected

6 Carnegy, H., 'France to Set Nuclear Power Cap', *Financial Times*, 18 June 2014.

7 Maynard, M., 'Tesla Shows Signs It's Struggling with Manufacturing', *Forbes Magazine*, 11 May 2014, forbes.com/sites/michelinemaynard/2014/11/05/why-tesla-should-take-heat-for-missing-its-targets/

8 'McKinsey on Susatainability & Resource Productivity', McKinsey & Company, August 2014, mckinsey.com/client_service/sustainability/latest_thinking/mckinsey_on_sustainability

9 De Neve, P., 'Electric Vehicles in China', Belfer Center, Harvard, June 2014, belfercenter.ksg.harvard.edu/publication/24345/electric_vehicles_in_china.html

10 Zach, 'Electric Car Sales Growing Much Faster than Hybrid Sales Did', EVObsession, 15 October 2014, evobsession.com/electric-car-sales-growing-much-faster-hybrid-sales-chart/

11 Michael Weinhold, Siemens, pers. comm.

CHAPTER FOURTEEN

1 Melillo, J. M., Richmond, T., Yohe, G. (eds.), *Highlights of Climate Change Impacts in the United States: The Third National Climate Assessment*, US Global Change Research Program, 2014.

2 Ackerman, D., *The Human Age: The World Shaped by Us*, Norton, New York, 2014.

3 Vince, G., *Adventures in the Anthropocene: A Journey to the Heart of the Planet We Made*, Chatto & Windus, London, 2014.

4 Campra, P. *et al*., 'Surface Temperature Cooling Trends and Negative Radiative Forcing Due to Land Use Change towards Greenhouse Farming in Southern Spain', *Journal of Geophysical Research* 113, D18, 23 September 2008, onlinelibrary.wiley.com/doi/10.1029/2008JD009912/full

5 Grossman, D., 'With Sawdust and Paint, Locals Fight to Save Peru's Glaciers', *PRI's The World*, 25 September 2012, pri.org/stories/2012-09-25/sawdust-and-paint-locals-fight-save-perus-glaciers

6 news.wikinut.com/How-to-restore-glaciers-by-Eduardo-Gold/3lq-j7a_/

7 Grossman, D., 'With Sawdust and Paint, Locals Fight to Save Peru's Glaciers', *PRI's The World*, 25 September 2012.

CHAPTER FIFTEEN

1 Crutzen, P. J., 'The Possible Importance of COS for the Sulfate Layer of the Stratosphere', *Geophysical Research Letters* 3, 73–76, 1976.

2 Crutzen, P. J., 'Albedo Enhancement by Stratospheric Sulfur Injections: A Contribution to Resolve a Policy Dilemma?', *Climatic Change* 77, 3–4, 211–20, August 2006, link.springer.com/article/10.1007%2Fs10584-006-9101-y

3 McClellan, J. *et al*., 'Cost Analysis of Stratospheric Albedo Modification Delivery Systems', *Environmental Research Letters* 7, 3, 034019, 2012, iopscience.iop.org/1748-

9326/7/3/034019/article

4 Crutzen, P., 'Albedo Enhancement by Stratospheric Sulphur Injections: A Contribution to Resolve a Policy Dilemma?', *Climatic Change* 77, 3–4, 211–20, August 2006.

5 Teller, E. *et al.*, 'Global Warming and Ice Ages: I. Prospects for Physics Based Modulation of Global Change', Lawrence Livermore National Laboratory, Livermore, CA, USA, 1997.

Keith, D. W., 'Geoengineering the Climate: History and Prospect', *Annual Review of Energy & Environment* 25, 245–284, 2000.

6 Weier, J., 'On the Shoulders of Giants: John Martin (1935–1993)', NASA Earth Observatory, 10 July 2001.

7 Wallace, D. *et al.*, 'Ocean Fertilzation: A Scientific Summary for Policy Makers', UNESCO, 2010, unesdoc.unesco.org/images/0019/001906/190674e.pdf

8 Helmholtz Association, 'Lohafex Provides New Insights on Plankton Ecology', *EurekaAlert*, 29 March 2009, eurekalert.org/pub_releases/2009-03/haog-lpn032409.php

9 Convention on Biological Diversity, 16, 'Biodiversity and Climate Change', cbd.int/decision/cop/?id=11659

10 Tollefson, J., 'Ocean-fertilization Project off Canada Sparks Furore', *Nature* 490, 7412, 2012, nature.com/news/ocean-fertilization-project-off-canada-sparks-furore-1.11631

11 Moore, D., 'Ocean Fertilization Experiment Loses in B.C. Court; Charges Now Likely', *Globe and Mail*, Vancouver, 3 February 2014, theglobeandmail.com/news/british-columbia/ocean-fertilization-experiment-loses-in-bc-court-charges-now-likely/article16672031/

12 Zubrin, R., 'Geoengineering Could Turn Our Long-barren Oceans into a Bounty', *National Review*, 22 April 2014, nationalreview.com/article/376258/pacifics-salmon-are-back-thank-human-ingenuity-robert-zubrin

13 Barry, J. P. *et al.*, 'Effects of Direct Ocean CO_2 Injection on Deep-sea Meiofauna', *Journal of Oceanography* 60, 759–66, 2004, imedea.uib-csic.es/master/cambioglobal/Modulo_III_cod101608/Tema_8-acidificaci%C3%B3n/pH/J%20Oceanogr60pp759-766%20deep-sea.pdf

14 Keller, D. P. *et al.*, 'Potential Climate Engineering Effectiveness and Side Effects during a High Carbon Dioxide-Emission Scenario', *Nature Communications* 5, 3304, 2014, nature.com/ncomms/2014/140225/ncomms4304/full/ncomms4304.html

15 Shukman, D., 'Geo-engineering: Climate Fixes "Could Harm Billions"', BBC News, 26 November 2014, bbc.com/news/science-environment-30197085

16 Warwick, Politics and International Studies, Clare Heyward, Profile, www2.warwick.ac.uk/fac/soc/pais/people/heyward/

17 'Climate Intervention: Carbon Dioxide Removal and Reliable Sequestration', and 'Climate Intervention: Reflecting Sunlight to Cool Earth', National Academy of Sciences, 2015, nas-sites.org/americasclimatechoices/public-release-event-climate-intervention-reports/

18 'Climate Intervention: Reflecting Sunlight to Cool Earth', National Academy of Sciences, 2015.

19 *Ibid.*

20 'Climate Intervention: Carbon Dioxide Removal and Reliable Sequestration' and 'Climate Intervention: Reflecting Sunlight to Cool Earth', National Academy of Sciences, 2015.

21 *Ibid.*

22 Although it should be noted that the official definition of climate mitigation in the IPCC reports is '"Mitigation" is the effort to control the human sources of climate change and their cumulative impacts, notably the emission of greenhouse gases (GHGs) and other pollutants, such as black carbon particles, that also affect the planet's energy balance', mitigation also includes efforts to enhance the processes that remove GHGs from the atmosphere, known as sinks [Edenhofer, O. *et al.* (eds.),. IPCC Working Group, 'Climate Change 2014: Mitigation of Climate Change', IPCC *Working Group III Contribution to the Fifth Assessment Report of the Intergovernmental Panel on Climate Change,* Cambridge University Press, Cambridge and New York, 2014].

CHAPTER SIXTEEN

1 Lomax, G. *et al.*, 'Investing in Negative Emissions', *Nature Climate Change* 5, 498–500, 2015, nature.com/nclimate/journal/v5/n6/full/nclimate2627.html

2 *Climate Intervention: Carbon Dioxide Removal and Reliable Sequestration*, National Academies Press, unpublished, 2015, nap.edu/catalog/18805/climate-intervention-carbon-dioxide-removal-and-reliable-sequestration

3 Gimeniz, E. H., 'Agroecology and the Diasappearing Yield Gap', *Huffington Post* Blog, 12 September 2014, huffingtonpost.com/eric-holt-gimenez/agroecology-and-the-disappearing-yield-gap_b_6290982.html

4 Eagle, A. J. *et al.*, 'Greenhouse Gas Mitigation Potential of Agricultural Land Management in the United States: A Synthesis of the Literature', Nicholas Institute for Environmental Policy Solutions, Duke University, North Carolina, January 2012, nicholasinstitute.duke.edu/sites/default/files/publications/ni_r_10-04_3rd_edition.pdf

5 Smith, L. J. and Torn, M. S., 'Ecological Limits to Terrestrial Biological Carbon Dioxide Removal', *Climate Change* 118, 89–103, 2013.

6 Keller, D. P. *et al.*, 'Potential Climate Engineering Effectiveness and Side Effects during a High Carbon Dioxide-Emission Scenario', *Nature Comunications* 5, 3304, 25 February 2014, nature.com/ncomms/2014/140225/ncomms4304/full/ncomms4304.html

7 Canham, H. O., 'The Wood Chemical Industry in the Northeast: An Old Industry with New Possibilities', *Northern Woodlands*, 8 February, 2010, http://northernwoodlands.org/articles/article/the-wood-chemical-industry-in-the-northeast

8 Myers, F. D., *The Wood Chemical Industry in the Delaware Valley*, Ontario & Western Railway Historical Society, Middletown, New York, 1986.

9 Jirka, S. *et al.*, 'The State of the Biochar Industry 2013', International Biochar Initiative, 2014, biochar-international.org/State_of_industry_2013

10 *Ibid.*

11 'CNBC Disruptor 50', cnbc.com/id/101724522

12 N'Yeurt, A. *et al.*, 'Negative Carbon via Ocean Afforestation', *Process Safety and Environmental Protection* 90, 467–74, 2012.

13 *Ibid.*

CHAPTER SEVENTEEN

1 'Consumption by Fuel, 1965–2008', in *Statistical Review of World Energy 2009*, BP, 8 June 2009.

2 Power, I. M. *et al.*, 'Serpentinite Carbonation for CO_2 Sequestration, *Geoscience World* 9, 2, 115–21, 2013, elements.geoscienceworld.org/content/9/2/115.short

3 Amato, I., 'Green Cement: Concrete Solutions', *Nature* 494, 300–01, 21 February 2013, nature.com/news/green-cement-concrete-solutions-1.12460

4 Schuler, T., 'Driving Sustainable Innovation to Market: It Can't Just Be Green, It has to Be Better', Solidia Technologies, April 2015, solidiatech.com/wp-content/uploads/2015/04/SBM_Solidia-Technology.pdf

5 Armstrong, T., 'An Overview of Global Cement Sector Trends', *International Cement Review*, 2 September 2013, ficem.org/boletines/ct-2013/presentaciones2013/1-EXPERTOS/2_THOMAS-ARMSTRONG/ICR-FICEM-Presentation-Handout-30Aug13.pdf

6 newlight.com/

7 Schroder, T., 'Synthetic Photosynthesis—Turning Carbon Dioxide into Raw Materials', *Pictures of the Future*, Siemens, 19 December 2014, siemens.com/innovation/en/home/pictures-of-the-future/research-and-management/materials-science-and-processing-co2tovalue.html

8 Lomax, G. *et al.*, 'Reframing the Policy Approach to Greenhouse Gas Removal Technologies', *Energy Policy* 78: C, 125–136, 2015.

9 'Climate Intervention: Reflecting Sunlight to Cool Earth', National Academy of Sciences, 2012, nap.edu/catalog/18988/climate-intervention-reflecting-sunlight-to-cool-earth

CHAPTER EIGHTEEN

1 Stock, A., 'Australia's Electricity Sector: Ageing, Inefficient and Ill-prepared', Climate Council Australia, climatecouncil.org.au/australia-s-electricity-sector-ageing-inefficient-and-unprepared

2 *Ibid.*

3 *Ibid.*

4 'Southern Co Says Kemper Coal Plant Costs Still Climbing', Reuters, 28 January 2014, reuters.com/article/2014/01/29/utilities-southern-kemper-idUSL2N0L300U20140129

5 House, K. Z. *et al.*, 'Permanent Carbon Dioxide Storage in Deep-sea Sediments', *PNAS* 103, 12291–95, 2006, pnas.org/content/103/33/12291.full

6 Tohidi, B. *et al.*, 'CO_2 Hydrates Could Provide Secondary Safety Factor in Subsurface Sequestration of CO_2', *Journal of Environmental Science and Technology* 44, 1509–14, 2010.
7 *Ibid.*
8 *Ibid.*
9 Agee, E. *et al.*, 'CO_2 Snow Deposition in Antarctica to Curtail Anthropogenic Global Warming', American Meteorological Society, 2013.
10 Virgin Earth Challenge Team, pers. comm.

CHAPTER NINETEEN

1 'Two Degrees of Separation: Ambition and Reality—Low Carbon Economy Index 2014', PricewaterhouseCoopers, 2014, pwc.co.uk/assets/pdf/low-carbon-economy-index-2014.pdf
2 Beckman, K., 'Interview Fatih Birol: "Gradual change will not save us"', *Energy Post*, 3 June 2014, energypost.eu/interview-fatih-birol-iea-gradual-change-will-save-us/
3 *Ibid.*
4 *Ibid.*
5 Caldecott, B. *et al.*, 'Stranded Assets and Subcritical Coal: The Risk to Companies and Investors', Smith School of Enterprise and the Environment, Oxford University, 2015, www.smithschool.ox.ac.uk/research-programmes/stranded-assets/SAP%20Report%20Printed%20Subcritical%20Coal%20Final%20mid-res.pdf
6 Wood, E., 'Obama's New Carbon Plan Makes History for Clean Energy', Renewable Energy, 2 June, 2014, renewableenergyworld.com/rea/news/article/2014/06/obamas-new-carbon-plan-makes-history-for-clean-energy
7 Wood, E., 'Obama's International Climate Strategy: More Grease for Renewables', Renewable Energy, 10 September, 2014, renewableenergyworld.com/rea/news/article/2014/09/obamas-international-climate-strategy-more-grease-for-renewables?cmpid=rss
8 Brandt, A. R. *et al.*, 'Methane Leaks from North American Natural Gas Systems', *Science* 343, 6172, 733–35, 14 February 2014, sciencemag.org/content/343/6172/733.short?related-urls=yes&legid=sci;343/6172/733
9 Davenport, C., 'Obama Is Planning New Rules on Oil and Gas Industry's Methane Emissions', *New York Times*, 13 January 2015, nytimes.com/2015/01/14/us/politics/obama-administration-to-unveil-plans-to-cut-methane-emissions.html?hp&action=click&pgtype=Homepage&module=first-column-region®ion=top-news&WT.nav=top-news
10 'EU Leaders Agree CO_2 Emissions Cut', BBC News, 24 October 2014, bbc.co.uk/news/world-europe-29751064
11 'The US–China Climate Deal—By the Numbers', Climate Action Tracker, 12 November 2014, climateactiontracker.org/assets/publications/briefing_papers/CAT_release_20141112Final.pdf
12 *Ibid.*
13 'UN Members Agree Deal at Lima Climate Talks', BBC News, 14 December 2014,

bbc.com/news/science-environment-30468048

14 McGrath, M., 'Stern Warning: Legally-binding Climate Deal "Not Necessary"', BBC News, 8 December 2014, bbc.com/news/science-environment-30373738

15 Wood, E. 'Obama's International Climate Strategy: More Grease for Renewables', Renewable Energy, 10 September 2014.

16 Chen, K. and Reklev, S., 'China's National Carbon Market to Start in 2016—Official', Reuters, 31 August 2014, uk.reuters.com/article/2014/08/31/china-carbontrading-idUKL3N0R107420140831

17 Stock, A. *et al.*, 'Lagging Behind: Australia and the Global Response to Climate Change', Australian Climate Council, 3 November 2014, climatecouncil.org.au/globalresponsereport

18 Cama, T., 'China to Launch Carbon Trading Scheme in 2016', The Hill, 31 August 2013, thehill.com/policy/energy-environment/216340-china-to-launch-carbon-trading-system-in-2016

19 Höhne, N. *et al.*, 'China and the US: How Does Their Climate Action Compare?', Climate Action Tracker, 21 October, 2014, climateactiontracker.org/assets/publications/briefing_papers/CAT_briefing_China_and_the_US__how_does_their_climate_action_compare.pdf

20 Bannerjee, R., 'Coal-based Electricity Generation in India', *Corner Stone*, 11 April 2014, cornerstonemag.net/coal-based-electricity-generation-in-india/

21 windpowerindia.com/index.php

22 Chadha, M., 'India's Solar Power Capacity Tops 2600 MW', Clean Technica, 8 April 2014, cleantechnica.com/2014/04/08/indias-solar-power-capacity-tops-2600-mw/

23 Katakey, R. and Watanabe, C., 'India Gets Obama's Backing for $160 Billion Solar Push', *Bloomberg Business*, 27 January 2015, bloomberg.com/news/articles/2015-01-26/india-gets-obama-s-backing-for-160-billion-solar-push

24 Das, K. N., 'Obama Backs India's Solar Goals, Seeks Support for Climate Talks', Reuters, 25 January 2015, in.reuters.com/article/2015/01/25/india-obama-climatechange-idINKBN0KY0QN20150125

25 Landberg, R. and Pearson, N. O., 'Modi Signals Indian Shift toward Global Deal on Climate Change', *Bloomberg Business*, 26 January 2015, bloomberg.com/news/articles/2015-01-25/modi-shifts-on-climate-change-with-india-renewables-goal

26 Buckley, T., 'India's Plan to Stop Importing Coal Deals another Blow to Australia', RenewEconomy, 13 November 2014, reneweconomy.com.au/2014/indias-plan-stop-importing-coal-deals-another-blow-australia-68894

27 Katakey, R. and Chakraborty, D., 'Modi to Use Solar to Bring Power to Every Home by 2019', *Bloomberg Business*, 19 May 2014, bloomberg.com/news/2014-05-19/modi-to-use-solar-to-bring-power-to-every-home-by-2019.html

28 Rose, C., 'Wind Power in Africa to Increase Ten Times Over', European Wind Energy Association, 30 May 2013, ewea.org/blog/2013/05/wind-power-in-africa-to-increase-ten-times-over

29 Desertec-UK: Clean Power from Deserts, trec-uk.org.uk/

30 Satriastanti, F., 'Al Gore Praises Yudhoyono, Cites Indonesia's Geothermal Potential', *Jakarta Globe*, 9 January 2011, thejakartaglobe.beritasatu.com/archive/al-gore-praises-yudhoyono-cites-indonesias-geothermal-potential/

31 'India's 1st Geothermal Power Plant to Come up in Chhattisgarh', *Economic Times*, 17 February 2013, articles.economictimes.indiatimes.com/2013-02-17/news/37144613_1_geothermal-energy-geothermal-power-plant-national-thermal-power-corporation

32 'Global Energy-Related Emissions of Carbon Dioxide Stalled in 2014', International Energy Agency, 13 March 2015, iea.org/newsroomandevents/news/2015/march/global-energy-related-emissions-of-carbon-dioxide-stalled-in-2014.html

33 Briggs, H., 'Global CO_2 Emissions "Stalled" in 2014', BBC News, 13 March 2015, bbc.com/news/science-environment-31872460

CHAPTER TWENTY

1 Weaver, A. J. *et al.*, 'Long Term Climate Implications of 2050 Emission Reduction Targets', *Geophysical Research Letters* 34, L19703, 6 October 2007, garnautreview.org.au/CA25734E0016A131/WebObj/D07119780SubmissionGreenleapStrategicInstitute/$File/D07%20119780%20Submission%20Greenleap%20Strategic%20Institute.pdf

2 'Pathways to Deep Decarbonization', IDDRI, Sustainable Development Solutions Network, 2014, unsdsn.org/wp-content/uploads/2014/09/DDPP_2014_report_executive_summary_digit.pdf

3 'World Energy Outlook 2014 Factsheet', International Energy Agency, 12 November 2014, worldenergyoutlook.org/media/weowebsite/2014/141112_WEO_FactSheets.pdf

4 McKibben, B., 'The IPCC Is Stern on Climate Change—but It Still Underestimates the Situation', *Guardian*, 2 November 2014, theguardian.com/environment/2014/nov/02/ipcc-climate-change-carbon-emissions-underestimates-situation-fossil-fuels

5 Hansen, J. *et al.*, 'Target Atmospheric CO_2: Where Should Humanity Aim?', Columbia University, 2008, columbia.edu/~jeh1/2008/TargetCO_2_20080407.pdf

CHAPTER TWENTY-ONE

1 'Adani's Galilee Coal Mine Faces Fresh Legal Challenge', *Sydney Morning Herald*, 16 January 2015, smh.com.au/environment/adanis-galilee-coal-mine-faces-fresh-legal-challenge-20150115-12rg17.html

2 'Indian Miner Adani Condemns "Fascinating" Court Case Challenging Carmichael Coal Mine Approval', ABC Rural, 16 January 2015, abc.net.au/news/2015-01-16/court-action-by-environmental-activists-against-coal-mine/6021874

3 Wood, M. C., *Nature's Trust: Environmental Law for a New Ecological Age,* Cambridge University Press, New York, 2013.

4 Democker, M., 'Does the Public Trust Doctrine that Protects Air, Water, and Endangered Species Apply to Climate?', *Oregon Quarterly*, University of Oregon, 2014, oregonquarterly.com/natural-law

Index